T0205516

Approaches, Opportunities, and Challenges for
Eco-design 4.0

Samira Keivanpour

Approaches, Opportunities, and Challenges for Eco-design 4.0

A Concise Guide for Practitioners and Students

 Springer

Samira Keivanpour
Polytechnique Montréal
Montreal, QC, Canada

ISBN 978-3-030-87373-8   ISBN 978-3-030-87371-4   (eBook)
https://doi.org/10.1007/978-3-030-87371-4

This Springer imprint is published by the registered company Springer Nature Switzerland AG
The registered company address is: Gewerbestrasse 11, 6330 Cham, Switzerland

For my wonderful parents, Sadegh and Zahra

Preface

Integrating sustainability into the engineering curriculum is essential. In some engineering schools, the students should take a course in sustainable development at the beginning of the program. The motivation for writing this book came from the author's experience in teaching sustainable production courses at graduate and undergraduate levels. Different students from industrial, electrical, mechanical, and chemical engineering backgrounds could take this course to be familiar with the operationalization of sustainable development. Design for environment is one of the core concepts in sustainable production. The students are interested in this topic as it discusses the intersection of sustainability, technology, engineering design, and innovation. It stimulates creativity and encouraging discussion in the class and the students could share their ideas in different multidisciplinary teams. It reinforces innovative solutions and "out of the box" approaches to the problems. One of the new discussion topics on design for environment for engineers is the implications of new technologies for facilitating design practices. Industry 4.0 as the revolutionary paradigm in industry and consumer-centric products are two essential trends that influence concurrent engineering. The integration of sustainability into these new trends builds an innovative space of design in the future. This book aims to discuss the integration of sustainability, industry 4.0, and consumer-centric products to highlight the existing gaps in the literature and provide the conceptual frameworks and the perspectives of the applications.

Today, developing tools and methods for advanced management of the product through their life cycle in the new paradigms of sustainability, manufacturing, and information technology is crucial. The new trends in manufacturing, information technology, and logistics influence the circular economy. The long-term goal is to provide the recycling companies, manufacturers, and the other key players in the life cycle management of the complex products with tools and methods for analyzing the implications of these new trends and support the robust circular economy. The new paradigm of Industry 4.0 affects the green supply chain includes green design, green manufacturing and remanufacturing, reverse logistics, and waste management. The role of customers is changed in the new paradigm of manufacturing. They could involve in design and contribute to the life cycle of the products.

Mass individualization provides opportunities for small companies to contribute to the design and manufacturing of the products. This paradigm causes multiple companies and buyers involved in product design. We will have more open platform products that facilitate the integration of computer-embedded systems and mechanical modules. In this book, we will address the opportunities of the implications of cyber-physical systems, big data analytics, Internet of Things, additive manufacturing, and simulation in the following areas in the eco-design context:

- Selecting low impact materials
- Choosing manufacturing processes with environmental considerations
- End-of-life strategies
- Applying design approaches for disassembly
- Cocreation design and integrating stakeholders values into the design

Several illustrations are provided to summarize the key concepts in different sections for facilitating comprehension. Moreover, the experience of the author in exchange with eco-design teams is integrated into discussions to highlight the requirements of designers in developing future eco-design tools. Hence, the junior researchers could use this discussion in developing the methods and techniques in the context of design for environment. The book includes five chapters. Chapter 1 focuses on the customer-centric paradigm, new trends in product design, and proposing a conceptual framework for eco-design in this new context. Chapter 2 discusses the detailed applications of Industry 4.0 technologies from a product, material, and process perspective. Chapter 3 provides design for EoL and enhancing the methods and techniques with digital technologies. Chapter 4 discusses the contributions of lean and industry 4.0 in circular design and finally, Chapter 5 provides the application perspectives in the aircraft eco-design practices.

Montreal, QC, Canada   Samira Keivanpour
June 2021

Contents

List of Figures

List of Tables

List of Abbreviations

3R	Reducing, Reusing, Recycling
AI	Artificial Intelligence
AR	Augmented Reality
BoM	Bill of Materials
CAD	Computer-Aided Design
DfD	Design for Disassembly
DfE	Design for Environment
DfX	Design for Excellence
EoL	End of Life
IoT	Internet of Things
LCA	Life Cycle Assessment
LCC	Life Cycle Cost
MCDM	Multi-Criteria Decision-Making
MIS	Management Information System
NST	Natural Step Theory
OEM	Original Equipment Manufacturer
OLAP	Online Analytical Processing
PLM	Product Life cycle Management
PSS	Product, Service, Systems
SDG	Sustainable Development Goals
S-LCA	Social Life Cycle Assessment
SME	Small Medium Enterprise
UN	United Nations
VR	Virtual Reality
WEEE	Waste Electrical and Electronic Equipment recycling

Chapter 1
Design for Environment in Consumer-Centric Paradigm

1.1 Introduction

Design for environment (DfE) is integrating sustainability into products and processes during their whole life cycle. As it includes the solutions for reducing the product's footprint at the source, it is considered the most effective solution for respecting sustainable development goals. As this field is a multidisciplinary area, different expertise should be integrated for developing the tools and applying the practices.

In recent years, some essential trends influenced product design. The first is the role of customers in the life cycle of the products. Consumers are active players from design to EoL of the product's lifetime. At the design stage, they prefer to collaborate to customize the products based on their preferences. During the use phase, they prefer customized service solutions, and at EoL, they are more environmentally responsible and active players in product take-back and finding sustainable end-of-life solutions. The second trend is sustainability. Consumers and manufacturers are more aware of the products' environmental and social impacts and seek effective solutions. The third trend is Industry 4.0 and digital technologies that transformed the product's life cycle and its value chain. Despite evolving the literature on technology innovation and new product development, few studies focused on the implications of these new trends on DfE. This chapter focuses on the impacts of the consumer-centric paradigm on DfE. The recent literature on this topic is provided and a conceptual framework based on the principles of concurrent engineering is proposed. The application example from the footwear industry is discussed to highlight the elements of this framework. The rest of the chapter is organized as follows: Sect. 1.2 provides some introduction about DfE and the new trends. Section 1.3 provides an overview of the customization and collaborative design, the new approaches in product development, and the design challenges. Section 1.4

proposes a conceptual framework, Sect. 1.5 discusses the application perspective, and finally, the conclusion and future research are provided in Sect. 1.6.

1.2   DfE and Evolution in Product Design and Technology

According to natural step theory (NST), increasing the demand for natural resources and depleting these resources without recovery solutions could be considered as a funnel metaphor of sustainability and shows how we put pressure on the planet with this unsustainable behavior for several years. The only way to get out of this funnel and decreasing the pressure is to sustain the demand/supply and helping the planet for the natural recovery of the depleted resources (Natural Step Theory n.d.).

The key targets for operationalization of sustainable development are specified in the UN sustainable development goals blueprint that highlight the priorities for taking actions for a sustainable world (UN-website). Among these goals, goal 12 focuses on sustainable production and consumption. It includes the sustainable use of natural resources, waste management, applying sustainability through 3R, and management of the product's life cycle. Hence, DfE could be defined based on the sustainability of products and processes considering all stages of the product life cycle to decreasing the environmental impacts, increasing the sustainable use of resources, and decreasing the risks for human health and safety (Fig. 1.1).

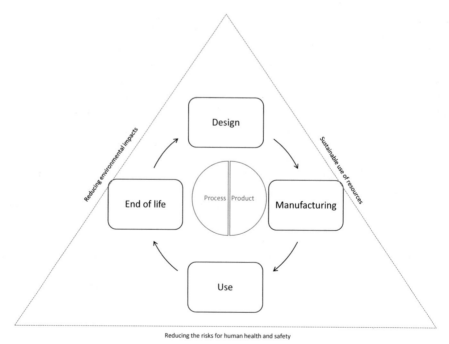

Fig. 1.1 DfE definition

In the new paradigm of sustainability, the circular economy is defined as designing the systems in a way to m inimize the resources, wastes, emissions, and energy leakage by long-lasting design, maintenance, and via 3R (reducing, reusing, and recycling) approaches (Geissdoerfer et al. 2017). This proactive approach to sustainability should be integrated into new paradigms in manufacturing, emerging technologies, and new trends in business.

Figure 1.2 shows the top trends in sustainability since 2010. In the 2010s, we had more corporate social responsibility awareness, and in 2015 with UN goals, a global active participation for sustainability occurred. In the 2020s with Industry 4.0 technologies, we have destructive technologies in the context of Industry 4.0 including the Internet of things, big data, artificial intelligence, and 3D printing. These technologies changed the forward supply chain radically. Moreover, they also affect the green supply chain, reverse logistics, and waste management. Hence, the existing trend is how these technologies could improve recovery networks and enhance sustainability performance. The next paradigm is realizing Society 5.0 with a balance between technology revolutions and creating value for all stakeholders and the ecosystem of players. According to Fukuda (2020), the future of science, technology, and innovation include the rise in data-driven innovation. The data-driven approaches include data collection, data analysis, and decision-making in different cycles that involve value creation feedback. Big data facilitates the interaction between manufacturers, consumers, and communities. However, this evolution involves different socioeconomic risks such as labor, capital, and spatial risk. Hence, Society 5.0 provides solutions for these risks via collaboration between humans and robots, creating values in the ecosystem of players and considering all stakeholders, and making the balance between the global and the local sphere too. Hence, empowering small, medium enterprises (SMEs) and creating values for communities is essential.

According to Oliveira et al. (2019), nowadays, consumers are more knowledgeable about different dimensions of the products including reliability, quality, and technical specification. They need to play an active role in product development.

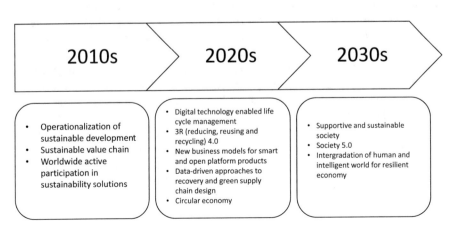

Fig. 1.2 Top trends in sustainability

Codesign and co-creation of personalized products are an essential part of this experience. The authors explained that mass customization and codesign are related concepts (Ulrich et al. 2003; Gilmore and Pine 1997). The consumer-centric paradigm refers to the customization of the product or service based on the consumers' demand and considering the necessities for manufacturers to respond to the required flexibility (Fogliatto et al. 2012).

There are a couple of challenges for DfE. This field is multidisciplinary areas and different expertise should be integrated to develop the solutions. The life cycle of products and the requirement for visibility in terms of data for each stage of this life cycle are also challenging. Addressing all environmental and social impacts of the products and processes during the whole life cycle is a complex task in comparison to focusing on regular supply chain challenges. Fiksel (2009) also addressed these challenges and discussed that despite a lot of examples of DfE in different enterprises, this field is an opportunistic area for researchers and practitioners, and more systematic approaches and techniques are required to fill these gaps. Hence, different aspects of innovation, technologies, engineering practices, and social and environmental impacts make DfE a challenging area.

Now, let us look at these complexities from the new paradigms' lenses too. The customer-centric paradigm is a redefinition of perceived value for the customers and involving more key players in the design stage. It increases the complexity of communication and dealing with the multidisciplinary nature of the eco-design field. Evolution in product life cycle management and using data-driven approaches facilitates visibility and tracing the product's data through its lifetime. Furthermore, close relationship with customers during different stages of business and the physical life cycle leads to more sustainable products. Table 1.1 summarizes the implications of customization, sustainability, and novel technologies in different phases of the products' life cycle.

Fiksel (2009) discussed four principles strategies of DfE. These four strategies are designed for dematerialization, design for detoxification, design for revalorization, and design for capital protection and renewal. The first strategy focuses on minimizing the energy and resource consumption in different stages of the life cycle of products. This strategy aims to reduce the wastes at the design stage and increase the durability of the product for a longer lifetime. The second strategy focuses on minimizing hazardous materials and introducing cleaner technologies. Design for valorization is much more considered in the circular economy context for facilitating reusing, recycling, and recovery options of the products. Hence, investment in these solutions at the design stage could be aligned with circular economy objectives, and the induced complexities should be managed to achieve a win-win strategy. Design for capital protection focuses on the balance between the used resources and generating and renewing different types of capital (human, natural, and economic) for sustaining productivity. These strategies could be mapped in a space of implications of customer-centric paradigm and Industry 4.0 development (see

Table 1.1 The implications of three trends of customization, sustainability, and Industry 4.0 on product life cycle

Products life cycle	Customer-centric implications	Sustainability implication	Industry 4.0 implications
Design	Codesign Adapting with changes The need for an interactive interface	Applying life cycle analysis for the sustainability of products and process	Advanced tools and simulations Cloud-based design for facilitating codesign and collaborative product platform
Manufacturing	Dealing with the complexity of production More flexibility and agility	Reducing the wastes Better use of resources More coordination with suppliers for eco-efficient solutions	Increasing efficiency of supply chain Better traceability Value creation through advanced transmitting and analyzing data tools and vertical and horizontal integration in supply chain
Use	The need for a long-lasting relationship with products Service solutions	Sustainability of maintenance Increasing product useful life	Smart products and better relationship between products and customers More safety Better traceability
Disposal	More customer awareness Better choice of materials for recycling and reusing More viable EoL solutions	Reducing hazardous materials Reusing, remanufacturing, and recycling options	Better life cycle management Enhancing disassembly, dismantling, and recovery solutions Better logistics of take-back and recovery

Fig. 1.3). Some of these eco-design solutions could be enhanced by these paradigms or face some challenges or even become less important. With novel technologies and personalization in the products, we have more focus on reducing and reusing options than recyclability. Reducing energy and resources in green design and using more natural and biomaterials reduce the concern of recyclability. With customization, consumers are also more knowledgeable about hazardous materials and they prefer using more natural sources for their health. The long-lasting relationship between products and consumers as the result of customization and connectivity technologies could reduce energy and resources usage. Product/service solution is also evolving with the customization and digital technologies development. Society 5.0 will bring the balance between the depleted resources and regeneration of the capital for the sustainability of the communities and creating values for manufacturers and the consumers.

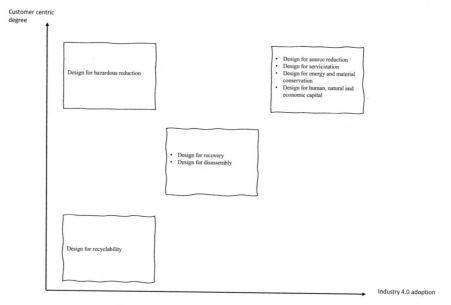

Fig. 1.3 Mapping the eco-design practices in consumer-centric and Industry 4.0 paradigms

1.3 Literature Review of New Trends in Products Design

To review the eco-design studies in customer-centric paradigm, we should consider the recent development in product design literature. For this aim, the new trends in product design including customized, smart, open-architecture products, integration of sustainability into the design of these products, and implications of digital technologies and Industry 4.0 should be considered. In this section, the recent publications that focused on these trends are reviewed to highlight the gaps and aid in extracting the elements of the proposed conceptual framework in this chapter.

1.3.1 Customer-Centric Paradigm and Collaborative Design

Turner et al. (2020) studied the impacts of codesign experience in mass customization on customers' satisfaction and loyalty. The authors proposed a conceptual framework to evaluate how codesign experience influences the perceived value such as complexity, control, and enjoyment and leads to satisfaction and finally the loyalty toward online mass customization. The authors concluded that firms should develop effective tools and interface for enhancing the collaborative design and codesign experience. Artificial intelligence and integrating real-time data could improve the codesign experience of customers. Developing a self-adapting interface in mass customization is valuable and enhances the relationship between customers and designers for better personalization.

Kumtepe et al. (2020) proposed a smart customization tool for a portable ramp for wheelchair users. The case study well presented a collaborative design tool that provides the interaction of users at the design stage. The interface should interact with users during the customization process. The authors discussed that the new paradigm in the manufacturing of smart customized products provides the personalized user manual during the customization process as the result of the users' involvement. The authors explained that in the future with 3D printing technology, the design tools will be offered to customers in the place of the real products. There are some limitations in this collaborative design: the complexity of geometry and plates of composite material makes the challenges in involving the consumers in the material selection phase. The user-friendly interface could enhance consumers' experience. The material mechanical library could be integrated into the collaborative design to facilitate the material selection for the customers. 3D design tools could also facilitate users' experience.

Elgammal et al. (2017) proposed a collaborative approach to product-service design and customization. In this model-driven approach, the authors considered two models: product configurator that is independent of the platforms and includes customers' requirements and engineers' consideration. 3D modeling and visualization could also be integrated into this part. The second model is a master plan includes suppliers' products and services and quality assurance blueprints.

Song and Sakao (2017)* developed a framework for the customization of product/service systems. The authors explained that increasing competition, customers' demand variety, and sustainability force the manufactures to provide service solutions to the customers. There are some challenges in designing product-service systems. First, the involvement of different stakeholders could make conflicts in different design features. Second, the customers' needs are dynamic during the use phase of the product, and these changes should be integrated into designing the service solutions for the customers. Hence, a systematic approach is required to customize the product-service systems based on these characteristics.

1.3.2   A Trade-off Between Flexibility and Complexity of the Design of the Customized Product

Zawadzki and Zywicki (2016) explained the integration of additive manufacturing and augmented reality into smart design and production control. The authors discussed the role of automation and flexibility in quick response strategy and designing mass customized products. The fast response to the clients' requests requires a high level of product design automation and rapid manufacturing tools as well as a high level of flexibility in manufacturing systems. The authors also explained the knowledge-based engineering systems that are required for designing digital product models. According to these authors, virtual reality facilitates the verification process of product design and comparison of different options. Hybrid prototyping is the integration of rapid prototyping and virtual prototyping and could accelerate the testing and verification process.

Modrak and Soltysova (2020) proposed a method for assessing the complexity of mass customization and minimizing configuration conflicts. The authors explained that higher production costs and the challenges of the manufacturing systems are not the only difficulties related to customization. Configuration conflicts are a major problem with a variety of options offered to customers and should be avoided in mass customization. The authors of this work explained different types of complexity in the context of personalized products. The first type of complexity is related to the product structure, the number of parts, and interrelationships.

For customized products, we face internal and external complexities related to the manufacturers' and customers' perspectives, respectively. The internal complexity is related to the number of inputs, energy, and information that are entered into manufacturing systems. The external complexity is related to customers' views during product design stage. The goal is to decrease the negative complexity and avoid conflicts as a result of the variant options offered to customers at the design stage.

Mourtzis et al. (2018) also performed a study on product-service system complexity metrics in mass customization and the implications of Industry 4.0. They addressed the complexity via information content, the quantity and diversity of information. They considered the quantity of the products and services as a factor of complexity for personalized products. The variants of products and services and the feedback from the external sources related to the quantity also influence the complexity.

1.3.3 Product Design and New Trends in Novel Technologies, Customers' Preferences, and Sustainability

Medini et al. (2020) discussed the integration of sustainability into product customization and portfolio management. The authors addressed the research performed in the SUSTAIN project for studying the integration of sustainability in the practitioner's decisions. They highlighted the existing gaps in the mass customization literature review for addressing the sustainability implications of what is called now: mass customization for sustainability in EU projects (FP7 program).

Dissanayake (2019) studied sustainability in mass customization in the fashion industry. The author discussed the collaboration between manufacturers and customers during the design stage of the products leads to more sustainable products. This could be achieved via the better relationship of customers and products as the result of the codesign, reducing the wastes due to using the customized materials, and the customers' satisfaction, using the local resources rather than the global supply chain, enabling eco-friendly printing technology and the possibility of reusing and recycling.

Yan and Chiou (2020) studied digital customization in the clothing industry. The interactive digital technology provides the opportunity for the customers to craft their products and collaborate at the design stage. They conducted the literature review as well as the interview with customers to identify the important indicators in digital customization. Authenticity value that includes reliability, responsiveness, interactivity, and real-time experience is the top-ranked coefficient from the customers' perspective. The second dimension is social value and esthetics and utility are in third and fourth place. The authors concluded that collaborative customization corresponds to authenticity value and refers to creating the collaborative platform between the producer and customers for a different experience of designing the final products and meeting the requirements of both parties.

Trollman and Trollman (2019) performed a sustainability assessment of smart innovation in mass customization and digital manufacturing. The authors discussed the economic, workforce, and environmental aspects of mass production, mass customization, and digital manufacturing. They explained that the required flexibility in the manufacturing process for mass customization makes some challenges related to optimizing material and energy consumption. However, the traceability of the products and the take-back options for reusing and recycling and better EoL decisions could be the advantage of the personalized products. Providing service solutions for the customers and a long-lasting relationship between customers and the products could enhance the product life cycle performance. The modular design also improves recycling and reduces the complexity of disassembling. For digital manufacturing and using 3D printing, the research on sustainability is at the infancy stage. The design for environment is a great advantage of 3D printing. Providing spare parts and facilitating repairing and remanufacturing are the other benefits of 3D printing.

Godina et al. (2020) also studied the implications of additive manufacturing on the sustainability of the business models. They discussed the role of co-creation and the customers' contribution to the design stage. The new customer-centered business models with enabling Industry 4.0 technologies could evolve new business models. The customization also adds some challenges concerning product traceability, supply chain, and standardization in some industrial sectors such as aerospace and automobile manufacturing.

Li et al. (2020) studied the impacts of Industry 4.0 on the economic and environmental performance of the firms. The authors performed a survey in Chinese manufacturing firms. They concluded that digital technologies positively impact the environmental performance of firms. Digital technology improves the quality of information and processing capacity. The vertical and horizontal integration of the supply chain facilitates information exchange and the digital supply chain could improve environmental performance.

Chiu et al. (2020) provided a systematic approach to the production research literature. The authors highlighted the need for theoretical expansion in new product development, environmental solutions, and service value chain based on this systematic review, thematic analysis, and methodological advances. These areas are essential in the context of sustainable customized products.

Li et al. (2015) studied eco-design in electronics products and analyzed the past, present, and future trends. The authors presented the development of eco-design theories and applications since 1985. This work shows that the recent trends are more focused on the end of life, recycling, and recovery solutions as well as "disassembly for recycling." The authors emphasized that future studies should focus on "disassembly for remanufacturing." The opportunities and challenges related to new materials such as carbon nanomaterial and the Internet of things, developing the new business models and the new legislation, should be addressed.

1.3.4 Open Architecture Products Design and Adaptable Product Platform

Open architecture products are also a new paradigm in personalized product development. There is a fixed platform for the product, and the different modules could be added based on the consumers' preference. According to Koren et al. (2013), this type of product will take a large share of the economy in the future, and different SMEs (small-medium enterprises), as well as consumers could collaborate at the design stage of the products (Keivanpour 2020).

There is different taxonomy of new products based on the new trends in the market, globalization, sustainability, and innovative technologies. Zhang et al. (2015) used five categories for comparison of the characteristics of open-architecture products with the other types of products. These five groups are mass-produced products, mass-customized products, reconfigurable products, upgradable products, and open-architecture products. In another study, Zheng et al. (2020) used the classification based on three features, smartness, connectedness, and openness. The authors classified products into five groups: regular products without any smart or connectivity features, regular products with smartness features, regular products with connectivity capability, smart products, and open architecture, and finally, the smart, open architecture with connectivity features.

Chen et al. (2018) developed an optimization tool for the adaptable design of open-architecture products. Adaptable design is changing the functionality of the products via upgrading functional modules. These changes in features could be applied as the result of required changes for consumers during the life cycle of the products. Hence, the design process of open-architecture products includes adapting the functions and solutions based on the customers' needs and upgrading modules with an open interface for a new product configuration.

Levandowski et al. (2015) also introduced a model for designing an adaptable product platform and for engineering to order. This model could provide the variation in products platform considering the changes in the consumers' needs. It also leads to efficient development and designing of the products based on the progress in the design decisions. The model has two stages including module configuration in the first stage of the design and scalable configuration in the later stage for developing the product variant.

The synthesis in this literature highlights the following conclusions:

- Customers and key stakeholders' requirements should be integrated into the design stage of products in a dynamic and interactive way.

- Developing interactive tools that facilitate the adapting and upgrading of the product platform is essential.
- Codesign and collaborative product development are the new trend in personalized products development and should be enhanced by innovative technologies.
- The products-service solutions should be enhanced for sustainability and customers' satisfaction.
- The persevered customers' value should be revised and redefined in developing the new business models in a customer-centric paradigm.
- Smart, adaptable, connected, personalized, and open-architecture products are the new trends in product development, and integrating sustainability into the design of these types of products require further research.
- The implications of Industry 4.0, particularly the role of 3D printing, virtual and augmented reality, digital twin, and Internet of things, on design for environment of these types of products are not studied in the literature systematically.
- An integrated framework is needed to analyze the cross-section of these three trends of sustainability, Industry 4.0, and customer-centric products.

Figure 1.4 shows these highlighted points and the call for research in this context.

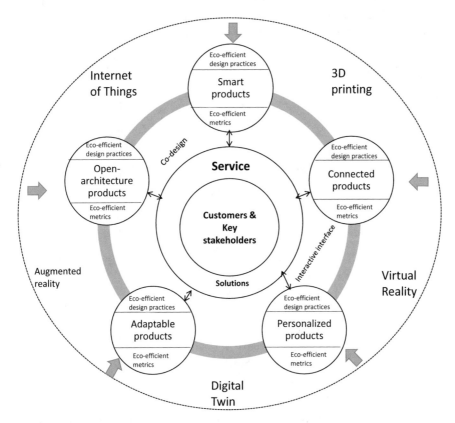

Fig. 1.4 Call for research in cross-section of Industry 4.0, customer-centric, and sustainability paradigms

1.4   Conceptual Framework

In this section, a conceptual framework is proposed to discuss the DfE in the new context of digital technologies and customization. Figure 1.5 shows this proposed framework for assessing the implications of the leverages for Design for X (DfX) and the outcomes. The building blocks of this framework are discussed, and application in the footwear industry is provided. The implications of Industry 4.0 will be discussed in detail in Chap. 2.

The principle of concurrent engineering includes applying design and engineering practices, using optimization tools for trade-off analysis, and updating the indicators and design features based on the customers' requirements. The design specifications could be updated during the life cycle of the product. In open-architecture products, the consumers could revise the specification at a later stage too. Therefore, considering the product life cycle is essential. Elhariri Essamlali et al. (2017) discussed the integration of the product life cycle into the collaborative design for smart products. The authors used the set-based concurrent engineering approach. In the traditional approach of current engineering (point-based), one alternative or solution is selected as the best solution. Hence, it is a time-consuming process to perform the trade-off analysis. With increasing the number of target features and the consumers' requirements, this process will be more complex and needs extensive assessment.

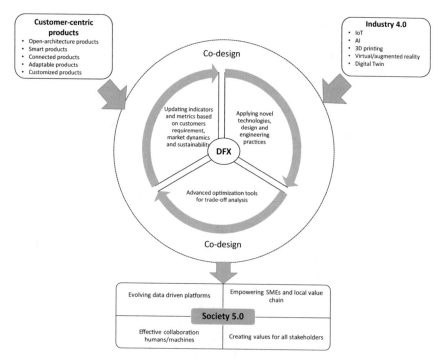

Fig. 1.5 The conceptual framework for DFX in new paradigms of sustainability, customer-centric products, and Industry 4.0

The set-based concurrent engineering was introduced as one of Toyota's principles in the product development system (Sobek II et al. 1999). The set-based approach uses multiple solutions in parallel and via a decision-making approach to use the feedback cycles for achieving the best solution. Therefore, the design principle is based on the optimization for mapping the design space, considering the constraints for finding feasible solutions, and validation and verification before applying the solution. In addition to the flexibility in design, a set-based approach could lead to decreasing the lead time and development costs and increasing the competitiveness and proficient of the design team (Raudberget 2010). In the customization paradigm, we could consider different sets of consumers' preferences, besides, to set of product design options and a set of manufacturable designs. The intersection between the different sets of possible options leads to the optimum solution. Figure 1.6 shows a set-based approach in the context of the customer-centric paradigm.

In the customer-centric paradigm, the design features could be modified during the life cycle and based on the customers' preferences. Industry 4.0 provides new design and engineering approaches and advanced optimization tools enhanced by data-driven approaches. CAD design could be integrated with virtual reality and augmented reality for facilitating trade-off analysis. 3D animation modeling of product modules could be integrated into the design tools for facilitating users' codesign experience.

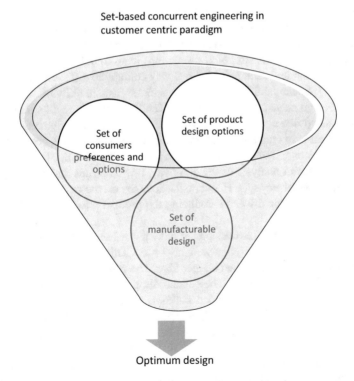

Fig. 1.6 Set-based concurrent engineering in the customer-centric context

According to Fiksel and Hayes-Roth (1993), the design requirement management process includes three steps: requirements analysis, requirement tracking, and requirement verification. Requirement analysis includes transplanting the customers' needs into the engineering design. The customization paradigm leads to increasing the complexity of the design. The preferences of customers during the product life cycle should be considered. The connectivity and smartness of the products require multilayers data analysis. Requirement traceability includes risk assessment and conflict management. Here, upgrading the information particularly for product/service solutions is needed. Industry 4.0 could facilitate this tracking during the design life cycle including design, virtual prototyping, engineering, and physical prototyping and testing. The validation and verification process needs compliance between the product requirements and the engineering constraints. A product digital twin could reduce the investment in this part and the design flaws risks. The digital twin provides a dynamic digital profile of a physical prototype and integrates a 3D model, historical, and real-time monitoring of the design process for optimizing the virtual reality.

1.5 Application Perspective

In this section, some evidence from the footwear industry is provided to highlight the elements of the proposed conceptual model. Nike is an excellent example of a manufacturer who worked effectively on DfE solutions in recent years. According to the manufacturer's website, Nike uses more than 16,000 materials in different products each year, and each pair of shoes may contain 30 different materials. Nike uses a Material Sustainability Index to compare the sustainability of different materials at the design stage and select the eco-friendly choices based on their environmental impacts (Nike website-1 n.d.). According to the chief design officer of Nike, the company aims to create and craft the future of the sport with a problem-solving approach in the context of circularity design (Nike -Video-1 n.d.). Based on the full value chain footprint analysis of the company, raw materials and production have major environmental impacts. Hence, Nike focused on the material choices at the design stage as a major driver for reducing the product's footprint (Nike website-3 n.d.).

One of the new customized designs of Nike is Air Force 1 Flyleather by Rouhan Wang, a Berlin-based artiste (Nike website-2 n.d.). This is a multicolored shoe with illustrative arts and 50% recycled leather fiber. The color, design, material, and special edition box are customized. The collaboration with artists and customers made a codesign experience. The advanced trade-off analysis tool provides an effective mechanism for selecting eco-friendly materials. The outcome could be the created value for all stakeholders, enhancing collaboration with local communities and encouraging the other customers to more sustainable and customized choices.

Some aspects could also be improved at the design stage. The first is improving the design. 3D design and virtual prototyping could decrease the time between developing the concept of the product and production (Bertola and Teunissen 2018; Papahristou and Bilalis 2017). Moreover, smart textile and intelligent wearable technologies are the next trend in the fashion industry. Data-driven platforms could aid in collecting and analyzing the large volume of customer data for a better relationship between users and manufacturers. The augmented reality and 3D simulation model could aid in designing the products based on the exact required shape and size for the customers. The connected products and the role of social media facilitates the integration of downstream and upstream of the supply chain. Hence, agility and flexibility to the customers' needs during the life cycle of the products and tracing the footprint for a more sustainable product could be realized. Figure 1.7 shows the essential elements of the new paradigm of DfE for a footwear product.

Figure 1.8 shows the codesign platform for designing a footwear product. Product life cycle data could be integrated into 3D animation modeling and virtual prototyping to provide a holistic approach for designing the product. The platform is shared between the design team and consumers for a better co-development experience. An interactive decision dashboard could be designed to aid consumers and the design team in the design life cycle. For example, eco-efficient metrics and eco-efficient practices could be evaluated for achieving more sustainable options. The eco-efficient metrics could involve energy usage, water usage, risk, source volume, and recovery/reuse. Eco-efficient practices could compare the possibility of different options such as material substitution, hazardous substance reduction, design for recycling, and product's life extension. Transferring this knowledge to consumers at the design stage could enhance their social and environmental awareness and lead

Customization
Trade off analysis for eco efficient material choice
Customers products' connection

Digital prototyping and 3D simulation
Advanced optimization tools for complex tradeoff
Personalization of shape and size
Shortening design cycle
flexibility in design

Connected and smart products
Horizontal and vertical integration in value chain
Better communities' involvement
Stakeholders' contribution

Fig. 1.7 The new paradigms in DfE of footwear products

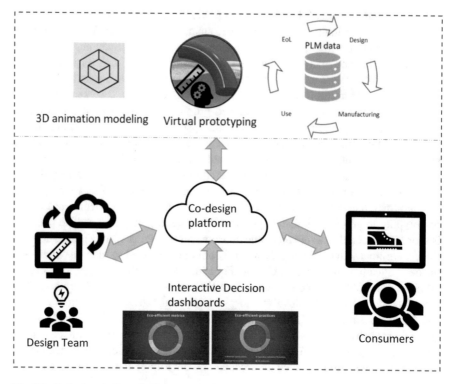

Fig. 1.8 Codesign platform for footwear product

to designing more sustainable products. Keivanpour and Ait Kadi (2018) discussed the role of visualization in DfE for analyzing the large volume of data and addressing the complexity and multi-attribute features. This visualization could facilitate the extraction of operative information and integrating different data including technical, functional, esthetic, and environmental. Moreover, it aids the decision-making process in a co-development platform.

1.6   Conclusion

This chapter discussed DfE in a new paradigm of customization and digital transformation. A conceptual framework based on concurrent engineering principles is proposed and application perspectives in the footwear industry are discussed. Integration of eco-design practices and eco-efficient metrics at the design stage for the new trends of product development including smart, connected, and open architecture products is a new research theme. More studies are required to investigate this integration via detailed case studies and application examples. The details of the implications of Industry 4.0 will be discussed in the next chapters.

References

P. Bertola, J. Teunissen, Fashion 4.0. Innovating fashion industry through digital transformation. Res. J. Textile Apparel **22**(4), 352–369 (2018)

Y. Chen, Q. Peng, P. Gu, Methods and tools for the optimal adaptable design of open-architecture products. Int. J. Adv. Manuf. Technol. **94**(1–4), 991–1008 (2018)

S.F. Chiu, L.E. Quezada, K.H. Tan, S.E.G. da Costa, Systemic approach to the new production research challenges. Int. J. Prod. Econ. **222**, 107495 (2020)

D.G.K. Dissanayake, Does mass customization enable sustainability in the fashion industry?, in *Fashion Industry—An Itinerary Between Feelings and Technology* (IntechOpen, 2019)

A. Elgammal, M. Papazoglou, B. Krämer, C. Constantinescu, Design for customization: A new paradigm for product-service system development. Procedia Cirp **64**(1), 345–350 (2017)

M.T. Elhariri Essamlali, A. Sekhari, A. Bouras, Product lifecycle management solution for collaborative development of Wearable Meta-Products using set-based concurrent engineering. Concurr. Eng. **25**(1), 41–52 (2017)

J. Fiksel, *Design for Environment: A Guide to Sustainable Product Development* (McGraw-Hill Education, New York, 2009)

J. Fiksel, F. Hayes-Roth, Computer-aided requirements management. Concurr. Eng. **1**(2), 83–92 (1993)

F.S. Fogliatto, G.J. Da Silveira, D. Borenstein, The mass customization decade: An updated review of the literature. Int. J. Prod. Econ. **138**(1), 14–25 (2012)

K. Fukuda, Science, technology and innovation ecosystem transformation toward society 5.0. Int. J. Prod. Econ. **220**, 107460 (2020)

M. Geissdoerfer, P. Savaget, N.M. Bocken, E.J. Hultink, The circular economy—a new sustainability paradigm? J. Clean. Prod. **143**, 757–768 (2017)

J.H. Gilmore, B.J. Pine, The four faces of mass customization. Harv. Bus. Rev. **75**(1), 91–102 (1997)

R. Godina, I. Ribeiro, F. Matos, B.T. Ferreira, H. Carvalho, P. Peças, Impact assessment of additive manufacturing on sustainable business models in Industry 4.0 context. Sustainability **12**(17), 7066 (2020)

S. Keivanpour, End of life management of complex products in an industry 4.0 driven and customer-centric paradigm: A research agenda. Accepted in MOSIM (2020)

S. Keivanpour, D. Ait Kadi, Strategic eco-design map of the complex products: Toward visualisation of the design for environment. Int. J. Prod. Res. **56**(24), 7296–7312 (2018)

Y. Koren, S.J. Hu, P. Gu, M. Shpitalni, Open-architecture products. CIRP Ann. **62**(2), 719–729 (2013)

E.D. Kumtepe, A.N. Başoğlu, E. Corbacioglu, T.U. Daim, A. Shaygan, A smart mass customization design tool: A case study of a portable ramp for wheelchair users. Heal. Technol. 1–15 (2020)

C.E. Levandowski, J.R. Jiao, H. Johannesson, A two-stage model of adaptable product platform for engineering-to-order configuration design. J. Eng. Des. **26**(7–9), 220–235 (2015)

J. Li, X. Zeng, A. Stevels, Ecodesign in consumer electronics: Past, present, and future. Crit. Rev. Environ. Sci. Technol. **45**(8), 840–860 (2015)

Y. Li, J. Dai, L. Cui, The impact of digital technologies on economic and environmental performance in the context of industry 4.0: A moderated mediation model. Int. J. Prod. Econ. **229**, 107777 (2020)

K. Medini, T. Wuest, D. Romero, V. Laforest, Integrating sustainability considerations into product variety and portfolio management. Procedia CIRP **93**, 605–609 (2020)

V. Modrak, Z. Soltysova, Management of product configuration conflicts to increase the sustainability of mass customization. Sustainability **12**(9), 3610 (2020)

D. Mourtzis, S. Fotia, N. Boli, P. Pittaro, Product-service system (PSS) complexity metrics within mass customization and industry 4.0 environment. Int. J. Adv. Manuf. Technol. **97**(1–4), 91–103 (2018)

Natural Step Theory, (n.d.), https://web.stanford.edu/class/me221/readings/NaturalStepOverview.pdf. Accessed 6 June 2021

Nike -Video-1, (n.d.), https://purpose.nike.com/innovating-sustainably

Nike website-1, (n.d.), https://purpose.nike.com/product-material-sustainability-indices

Nike website-2, (n.d.), https://www.nike.com/ph/launch/t/air-force-1-flyleather-rouhan-wang

Nike website-3, (n.d.), https://purpose.nike.com/value-chain-footprint

NSF, (n.d.), https://web.stanford.edu/class/me221/readings/NaturalStepOverview.pdf

N. Oliveira, J. Cunha, H. Carvalho, Co-design and mass customization in the Portuguese footwear cluster: An exploratory study. Procedia CIRP **84**, 923–929 (2019)

E. Papahristou, N. Bilalis, 3D virtual prototyping traces new avenues for fashion design and product development: A qualitative study. J. Textile Sci. Eng. **6**(297), 2 (2017)

D. Raudberget, Practical applications of set-based concurrent engineering in industry. J. Mech. Eng. **56**(11), 685–695 (2010)

D.K. Sobek II, A.C. Ward, J.K. Liker, Toyota's principles of set-based concurrent engineering. MIT Sloan Manag. Rev. **40**(2), 67 (1999)

W. Song, T. Sakao, A customization-oriented framework for design of sustainable product/service system. J. Clean. Prod. **140**, 1672–1685 (2017)

H. Trollman, F. Trollman, A sustainability assessment of smart innovations for mass production, mass customisation and direct digital manufacturing. In *Mass Production Processes*. (IntechOpen, 2019)

F. Turner, A. Merle, D. Gotteland, Enhancing consumer value of the co-design experience in mass customization. J. Bus. Res. **117**, 473–483 (2020)

P.V. Ulrich, L.J. Anderson-Connell, W. Wu, Consumer co-design of apparel for mass customization. J. Fashion Market. Manag. **7**(4), 398–412 (2003)

UN-Website., https://www.un.org/sustainabledevelopment/sustainable-development-goals/

W.J. Yan, S.C. Chiou, Dimensions of customer value for the development of digital customization in the clothing industry. Sustainability **12**(11), 4639 (2020)

P. Zawadzki, K. Żywicki, Smart product design and production control for effective mass customization in the industry 4.0 concept. Manag. Prod. Eng. Rev. **7**(3), 105–112 (2016)

J. Zhang, D. Xue, P. Gu, Adaptable design of open architecture products with robust performance. J. Eng. Des. **26**(1–3), 1–23 (2015)

P. Zheng, X. Xu, C.H. Chen, A data-driven cyber-physical approach for personalised smart, connected product co-development in a cloud-based environment. J. Intell. Manuf. **31**(1), 3–18 (2020)

Chapter 2
Industry 4.0 Impacts on the Sustainable Product, Process, and Material

2.1 Introduction

The studies on the implications of Industry 4.0 on sustainability are growing. Bonilla et al. (2018) focused on cyber-physical systems, the Internet of things, 3D printing, big data, and cloud to address these technologies deployment, relations to sustainable development goals, and their long term impacts on input and output flows of operation. In smart production, vertical and horizontal integration leads to the availability of real-time data related to material and energy flows. Big data analytics enhance efficiency and optimization and lead to better management of energy and material consumption. Additive manufacturing aids in fast prototyping and decreases waste and material flows. For goal 12 of SDG, digital technologies support increasing reliable operation and decreasing the material and waste flows. Oláh et al. (2020) also studied the impacts of Industry 4.0 on environmental sustainability. From customers' value proposition perspective, it leads to better customization and customer experience. In addition to cost saving, it facilitates new business models such as cloud-based, service-based, and process-oriented business models. Kamble et al. (2018) discussed the role of different digital technologies on process integration including human robots' collaboration and shop floor equipment integration. Hence, it improves safety, environmental impacts, and economic performance. The authors also addressed nine Industry 4.0 technologies and six principles including interoperability, virtualization, real-time capability, decentralization, modularity, and service orientation for supporting their sustainability implications. The number of research works that focused on sustainability and Industry 4.0 is increasing in recent years (e.g., Enyoghasi and Badurdeen 2021; Rojek et al. 2021; Ghobakhloo 2020; Bai et al. 2020; Bonilla et al. 2018).

S. Keivanpour, *Approaches, Opportunities, and Challenges for Eco-design 4.0*, https://doi.org/10.1007/978-3-030-87371-4_2

However, a review on its impacts on eco-design practices has not received much attention in the literature review. This chapter aims to address the role of selected Industry 4.0 technologies including data analytics, additive manufacturing, cloud, Internet of things, simulation/optimization, and augmented/virtual reality on eco-design practices. A review of trends, methods and tools, principles, and processes is required to highlight the role of technologies in the eco-design evolution path. Hence, a holistic approach is proposed for performing the literature review. Then, a framework is proposed to address the impacts of digital technologies on material, process, and product-based practices. The rest of this chapter is organized as follows: Sect. 2.2 addresses the literature review on eco-design. Sect. 2.3 proposes a framework for addressing the implications of the Industry 4.0 technologies on eco-design strategies from theories and practical points of view. Conclusion is provided in Sect. 2.4.

2.2   Literature Review

The eco-design research field is emerging. Several systematic literature reviews address the trends in strategies, practices, tools, and methods. In this section, a framework for synthesis in the literature is proposed to facilitate the discussion on the future of eco-design practices and the implications of Industry 4.0 technologies (see Fig. 2.1). Four streams including the review on practices and trends, the development of the methods and tools, the new paradigm of data-driven and product-service systems (PSS), and the organizational perspective are discussed.

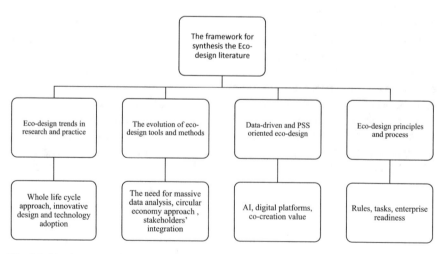

Fig. 2.1 The framework for literature review on DfE

2.2.1 Eco-Design Trends in Research and Practice

Several literature reviews have been conducted to highlight the future trends in DfE. In this section, some of these reviews are provided to link the role of Industry 4.0 in supporting these trends.

Bhamra (2004) discussed the eco-design trends in research and practices. The author divided the eco-design approaches into two categories of incremental and innovative approaches. The incremental approaches address the trends that focus on design improvement and optimization of existing products for decreasing the environmental impacts and cost-saving purposes. The innovative approaches address the radical changes in the design of the product considering the technology and social trends. The authors concluded that sustainable products in the future require more innovative approaches in the design of the product and service. They provided an example of the smart sink for eco-kitchen design to highlight the characteristics of the innovative eco-design and its multidisciplinary nature. Such radical changes require different sources of information including the environmental impacts of the products, life cycle analysis, efficiency targets at design, different case studies, and customers' preferences.

The first attempts of the enterprise in applying the eco-practices start with green design. These efforts are aligned with the idea of lean management for continuous improvement and efficient resource consumptions. According to Kane (2017), there are two models for a sustainable economy: eco-efficiency and ecosystem. Eco-efficiency aims to reduce the input resources for the optimum output. However, the rapid development of technologies encourages more consumption, and this is not aligned with the sustainable economy in the long term. The ecosystem model is designed based on nature's behavior to close the loops and consider the products during the whole life cycle. The green design could be considered in the context of eco-efficiency models. The next level of the enterprise efforts is considering the environmental impacts of the product during different stages of the life cycle and integrating the required solution into the early-stage design of the product. This ecosystem's point of view could improve the long-term sustainability objectives; however, it still needs to integrate the social aspects. Hence, the mature involvement of the enterprise in eco-oriented activities is sustainable design to consider three pillars of sustainability. The driving forces for eco-design could be addressed based on these three levels. In the first level, efficiency targets, competition, and mandatory legislation are key factors to adopt green practices. In the second level, the new trend in customers' preferences in eco-products stimulates innovative solutions and developing new products. The third level includes communications and integrating the values of all stakeholders at the design stage. These three evolving stages could be addressed in the developing trend of Industry 4.0 technology. The paradigm of factory 4.0 focuses on applying data-driven technologies and automation for operation. Industry 4.0 is the term that includes broader scope and technologies in product life cycle management and supply chain context. Society 5.0, as described in the previous chapter, is balancing between the advanced economic models and human

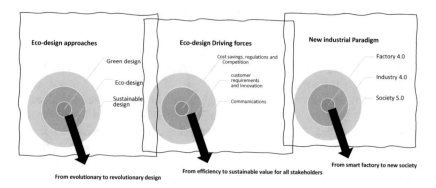

Fig. 2.2 The evolving trends in ecodesign, driving forces, and Industry 4.0

problems for achieving long-term sustainability. Hence, the future trend in eco-design includes development from efficiency to innovative sustainable solutions considering co-creation of the value for all stakeholders and implications of Industry 4.0 toward society 5.0 (see Fig. 2.2).

Kim et al. (2020) analyzed ten case studies in eco-design in three categories of EoL cases, use phase, and life cycle oriented. The authors emphasized the role of data analytics in sustainable design. They discussed the main challenges regarding the data during the life cycle of the product. These challenges could be classified into three parts: value chain evaluation that is related to the design, manufacturing, and distribution. The static data and too specific data during distribution are the challenges in this part. The second part is co-evaluation that is related to the use phase of the products and customers. The data are usually too generic or confidential in this part. The last part is related to EoL and recycling. Data tracing and availability at this stage is the main challenge. The authors discussed the research agenda based on the tools and methods in different eco-design case studies. In short term, it is needed to apply these tools in the industry. Hence, more practical research with real industrial applications is required. In the medium term, integration of the social dimension should be considered. In the long term, supplying data analytics and more holistic and system perspectives are required. The dynamics of the market and rapid changes in customers' preferences should be integrated into eco-design tools. Moreover, the sustainable design should be customized based on the product's systems and service-oriented approach. Business to business relationships and the possibility for traceability of data and information could enrich the eco-design practices.

Bhamra and Hernandez (2021) conducted a review on design for sustainability for explaining the trends, policies, and practices. The authors explained that one of the existing challenges in the eco-design approach is the product-oriented approach. There is a need to have a systemic view and consider the product-service system (PSS) approach. Integrating different actors in design, sharing concepts, and closing the loop are the other challenges. Focusing on innovation and considering the whole life cycle of the product is essential. One other recent theme in sustainability is the circular economy. Hence, developing the holistic approach, PSS and circular

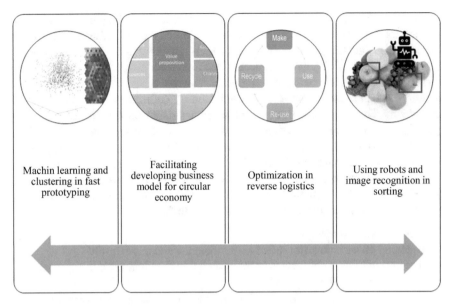

Fig. 2.3 AI role in the circular economy (based on Ellen MacArthur Foundation 2019 report cited in Bhamra and Hernandez 2021)

economy will form the new trends in design for sustainability. Artificial intelligence (AI) and circular economy as two main trends in the future influence eco-design research (Ellen MacArthur Foundation 2019). AI could impact sustainable products in different ways. First, with the possibility of prototyping and fast testing in the early stage of design. Second, facilitating designing sound business models for circular economy considering the uncertainties of demand and supply. Third, optimization of reverse logistics, and fourth, optimizing products take-back and recycling with image recognition and robots (see Fig. 2.3). In addition, the active role of the customer in product design is critical. Open-architecture products and open design and the need for the simplicity of the product architecture is the other trend in the eco-design context. The authors also linked the limitation in the application of the eco-design tools to the lack of data.

2.2.2   The Evolution of Eco-Design Tools and Methods

Lofthouse (2006) studied the challenges of eco-design tools for designers and addressed the requirements. According to this reference, the existing checklist is usually too general. There is no source to gather all information and data for designers including the case studies and examples from different industrial sectors. The process is time-consuming and more formal. Moreover, gathering all designers in a one-day workshop for mapping the eco-design process is challenging. Visual communication is required to facilitate the integration of the tools. Hence, the author introduced a holistic view to eco-design tools including guidance, information, visualization, simplicity, dynamic access, required training, and inspiration. Bovea and Pérez-Belis (2012) performed a literature review on the eco-design tools. The authors explained that these tools are not used by companies as a result of their complexity. They explained three essential characteristics for these tools: early integration, life cycle, and multi-criteria approaches. One of the challenges of the quantitative methods is the requirement of massive data at the early stage of product development. Integration of the eco-design tool in the late stage of the design will be limited to some minor changes. Pigosso et al. (2015) also conducted a literature review on the tools and methods of eco-design. The authors discussed the following points as the main future trend in the eco-design tools:

- PSS-oriented eco-design
- Focusing on sustainability rather than efficiency and eco-friendly
- Integrating LCA, LCC, and S-LCA, system thinking, and holistic approach
- Focusing on circular economy
- Considering all stakeholders and integrating managerial and strategic aspects

Rossi et al. (2016) reviewed the eco-design tools for addressing the barriers and effective industrial implementation. The authors classified the main barriers into three categories: (1) resources, time, the required staff, economic, as well as management of these resources, (2) the high number of tools and variety, and (3) the absence of multi-objective analysis. Faludi et al. (2020) proposed a roadmap in eco-design tools and methods. The authors emphasized that digital technology tools play an essential role in developing sustainable tools and methods and facilitate knowledge sharing and support industrial practices.

2.2.3 Data-Driven and PSS-Oriented Eco-Design

According to the synthesis in the literature of eco-design tools and trends, data-driven approach and PSS design are the main movements in this context. Feng et al. (2020) studied data-driven approaches in product design. The authors explained that data mining is vital for analyzing the implicit customers' requirements. This aids in finding the customers' preferences and meet their requirements. The product design history is also important, and text mining and data analytics tools could help designers. The other aspect is data sharing platforms for facilitating the collaboration among different actors on the product design stage including manufactures, customs, designers, and other stakeholders. Data-driven approaches also aid in addressing the uncertainties in optimization models. Different virtual design platforms based on virtual reality and digital 3D models are proposed for enhancing modeling capability and optimization. Karwasz and Trojanowska (2017) developed a CAD 3D approach in eco-design and disassembly sequence planning and design for recycling. The developed digital platform could aid in virtually disassembling the different components and optimize the design for disassembly. The authors used a database including joining techniques, compatible materials and assembly tools, and related recycling directives; then product structure, the weight of parts, and material composition will be analyzed to find recycling rate and optimum disassembly sequence. Trojanowska et al. (2017) applied virtual reality for design for disassembly. They created a 3D model in a CAD system and then inserted this model into a virtual reality environment. The simulation model is developed to support the 3D model. The disassembly tools databases will be added. Then, parts' disassembly time and disassembly/collision tool-part will be evaluated. The recycling rate will be calculated, and finally, the new variant of the product based on the modification will be proposed. Rojek and Dostatni (2019) developed an artificial neural network approach on material selection at the design stage for achieving the target recyclability rate of the product. Kim et al. (2021) developed an optimal modular-based family considering intellectual property and sustainability. The authors explained that one challenge in integrating the eco-friendly strategies at the design stage is the concerns related to patents and original equipment manufacturers. It usually reduces the commonality to prevent reuse. Today, blockchain technology is frequently proposed in the literature for the protection of intellectual property particularly in 3D printing design (e.g., Holland et al. 2018).

The research on PSS eco-design practices is under development. The role of stakeholders in co-creation value and the holistic approach is the highlighted points in the literature. da Costa Fernandes et al. (2020) performed a systematic review on PSS design approaches in a circular economy context. As the gaps in the literature, the authors mentioned that key performance indicators are not considered to measure the circularity of the value proposition. Moreover, shared value, the perceived values of different stakeholders, the complexity of their interaction and interdependency, and co-creation value have not received much attention in the literature. Dostatni et al. (2018a, b) discussed the innovative methods in the eco-design of home appliances. The author mentioned a holistic approach in eco-innovation considering innovation in services, products, procedures, and systems. Different methods such as simulation, brainstorming, crumbling, and play in words are suggested.

2.2.4 Eco-Design Principles and Process

The product development process and design tasks should be defined based on eco-design main rules, enterprise maturity in applying the practices and the requirements of the customers, and industrial technology leverage. Luttropp and Lagerstedt (2006) proposed ten golden rules for eco-design. These rules are related to the design, use phase, and end of life of the products. Pigosso et al. (2013) proposed a maturity model for eco-design. The authors introduced five capability level from incomplete to level five that addresses the continues improvement based on a performance evaluation framework. For integrating environmental consideration into the design stage, they also considered five levels from little experience on eco-design to fully incorporated level. Schischke et al. (2005) presented six phases in the product development process and the related eco-design activities. The first phase is planning. In this phase, the priorities in product design, the enterprise strategy, the existing environmental management system in the company, and the business environment should be addressed. In the conceptual phase, the integration of environmental aspects into the product specifications should be performed. The feasibility study, applying the checklists, and communication with suppliers should also be done in this phase. The third phase is detailed design. In this phase, the required eco-design tool and using certain databases lead to finding and selecting the appropriate design alternatives from material and structure perspectives. In the fourth phase, the proposed prototype should be tested and a benchmark study with existing products should be performed. Phase five includes launching the product and conducting related surveys and communication concerning quality, environmental impacts, and costs of products. The last phase is the product review for evaluating the performance of the developed product based on sustainability measures and applying the feedback for the next loops of the innovative design. This traditional product development process should be modified in the context of the customers' centric paradigm and digital technologies. Thomas et al. (2021) developed an eco-design approach based on the co-creation concept and consumers' participation in eco-design. The process includes the conceptual phase with a concept test with a small group of customers. Then, an interactive process is designed to integrate the customers' cocreation results and LCA analysis. Then, a validation step will be designed with a large group of consumers. Figure 2.4 shows the main steps of the

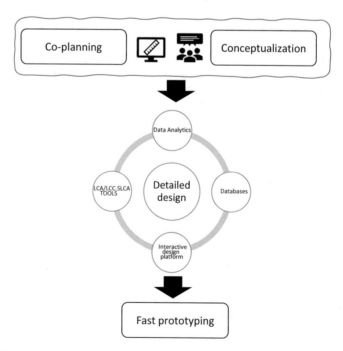

Fig. 2.4 The redefined product development process

redefined product development process and the related eco-design tasks in Industry 4.0 and the customer-centric paradigm.

2.3   The Implications of Industry 4.0 on Eco-Design Practices

The eco-design practices could be classified into three categories: those that consider the use phase, those that focus on the end of life stage, and the whole life cycle-oriented practices. In addition, the practices could be related to material, processes, and products. The sustainability design practices should be analyzed in different horizon plans. Hence, the framework illustrated in Fig. 2.5 is proposed to address the impacts of the selected Industry 4.0 technologies on eco-design.

Nine technologies are discussed in the literature of Industry 4.0. In this study, we focused on more relevant technologies in engineering design and sustainability. These technologies and their definitions are shown in Fig. 2.6 (for more information about Industry 4.0 in the circular economy, refer to Cagno et al. 2021).

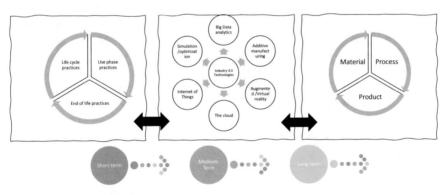

Fig. 2.5 The proposed framework for examining the impacts of Industry 4.0 on ecodesign practices

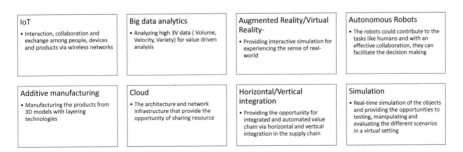

Fig. 2.6 The Industry 4.0 technologies discussed in this book

2.3.1 The Impacts on Sustainable Material Practices

Material-oriented practices play an essential role in eco-design. Design of the products is an integrated task, and multiple features and expertise should be considered simultaneously. For the material, in addition to functional and mechanical characteristics, recyclability and maintainability should be taken into account. The eco-design tools are developed first in the industries with more environmental legislation (automotive and electric/electronic). Now, more industrial sectors are involved in these practices. Moreover, small-medium enterprises (SMEs) are also interested in adopting eco-innovation solutions (Mathieux et al. 2007). Different practices could be integrated into the early stage of design to improve the eco-efficiency of used materials (Van Hemel and Cramer 2002; Borchardt et al. 2009). These practices are shown in Fig. 2.7. Several practices are related to the EoL phase including using recycled materials, facilitating recycling, and avoiding hazardous materials. Physical and chemical properties of the secondary stream of material are essential in the viability of these solutions. The uncertainties related to the quality and the price should be considered (Mathieux et al. 2007). Using data with quality enriched with data analytics tools in different product life cycle aids in dealing with these uncertainties.

Several examples illustrate the application of data mining, AI, cloud-based solution in eco-material practices. Cicconi (2020) proposed an interactive web-based platform as an eco-material tool. Integration of the recent technologies in developing digital mockups of the product and the consumers' preferences could stimulate innovative eco-material solutions. The author proposed a collaborative platform with codesign features for the integration of different users from designers to customers. A material selection tool is proposed to support the designers in testing and evaluating the material selection process. 3D models, different datasets, and reports are integrated into the platform for facilitating the evaluation process. A material

Fig. 2.7 The ecodesign practices (material perspective)—Created in Mindomo software

knowledge base platform is designed to support the sharing of data in different sub-processes including recycled materials, preprocessing, postprocessing, compounding, shaping process, and the posttreatment. The physical and virtual prototyping should be integrated into the collaborative platform. Physical conceptualization is required for testing processability and the quality of the recycled materials. The virtual platform with the simulation and optimization capability aids in the multi-attribute evaluation including the cost/benefits analysis, LCA, and conceptual mock-up for testing economic, environmental impacts, and functionality/appearance of the products. The results of physical and virtual prototyping will be shared via a collaborative platform with different players including manufacturers, suppliers, customers, and designers. Maximizing the use of standard and recyclable material facilitates recycling and sorting. The compatibility of materials based on different properties is also essential. Figure 2.8 shows the different material properties and some examples of the features in each group. Hence, multipotential compatibility analysis is required to consider the different properties' compatibility. Different AI techniques including decision trees, neural networks, machine learning, support vector machines are used for smart sorting and evaluating material compatibility at the design stage. Figure 2.9 shows some examples of eco-material practices using AI.

Scalice et al. (2009) discussed the challenges of material compatibility in polymer engineering for design for recycling. The authors explained that the wide range

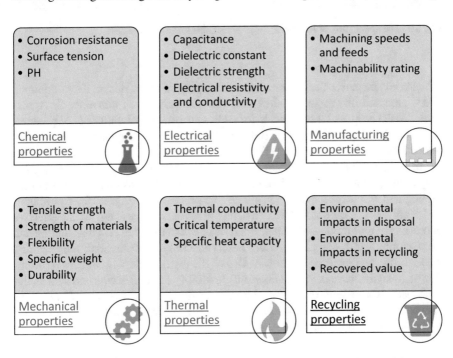

Fig. 2.8 Examples of different material properties that should be considered in material compatibility (see Rattan (2008) for more details)

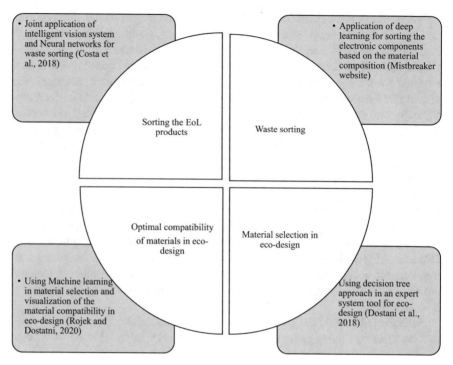

Fig. 2.9 The application of AI techniques in sustainable material practices

of materials increases the complexity of the adding process. Hence, they developed a new compatibility table for addressing different mixture of materials. de Aguiar et al. (2017) proposed a color scale for addressing the recyclability of the products at the design stage. For material indexes, they considered the hazardous material, the required technology for recycling, compatibility index, and contamination during the manufacturing process (for example, painting and welding). Hence, an integrated compatibility analysis is a complex and multidimensional study, and data mining and visualization could facilitate this process (see Fig. 2.10). Keivanpour and Kadi (2018) proposed OLAP (online analytical processing) as an effective approach for multidimensional data analysis of evaluating different dismantling and disassembling of a complex product based on the type of material, recyclability, replicability, and material scarcity. In OLAP, the data structure is cube form (multi-dimensional databases), and it allows the possibility of rapid access and data analysis based on a roll-up, drill-down, and dicing functions of the cube. Figure 2.11 shows the requirement for the pivot table and multidimensional data analysis for material compatibility in eco-design.

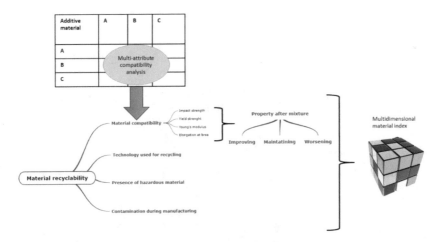

Fig. 2.10 Developing multidimensional material index based on the material compatibility and different recyclability features (an integrated index based on Scalice et al. 2009; de Aguiar et al. 2017)

2.3.2   The Impacts on Sustainable Process Practices

The sustainability of the manufacturing process is also critical. Different process-based practices enhance the sustainability of the products. These processes are shown in Fig. 2.12. Digital twin, big data analytics, and IoT are frequently cited in the literature as the enabling technologies for improving sustainability. The digital twin provides a dynamic digital model of a physical object and integrates a 3D model, historical, and real-time behavior for optimizing virtual reality. Rojek et al. (2021) discussed the application of the digital twin in the sustainability of the product life cycle. Data cleaning and data mining are important in building digital models. According to the authors, the data from several manufacturers are collected. These data are related to the companies with unit or small batch production. Hence, experience and knowledge in designing and manufacturing are vital. The real-time monitoring of the manufacturing process aids to reduce wastes during the machining process. The simulation could predict process disruption and improve the stability of the process. The maintenance tasks could be improved via simulation, optimization, and real-time monitoring. Zhang et al. (2017) proposed a big data analytics architecture for the sustainability of manufacturing and improving maintenance operations. Gu et al. (2017) applied IoT and big data analytics in improving the sustainability of the logistics of waste electrical and electronic equipment.

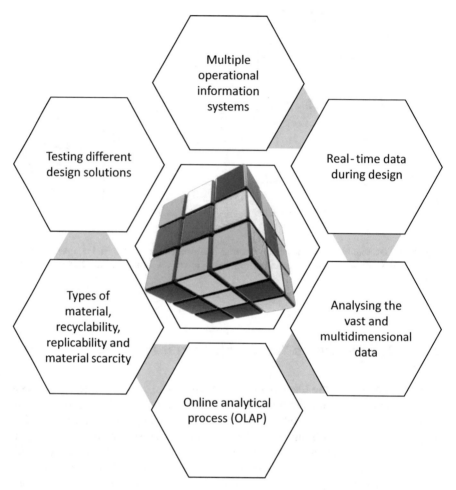

Fig. 2.11 The need for a multidimensional pivot table for assessing material compatibility

2.3.3 The Impacts on Sustainable Products Practices

The eco-innovative solutions and Industry 4.0 have changed the paradigms in product designs. These changes are related to the product structure, functionality, simplicity, and relation with customers. Figure 2.13 shows the common product eco-practices. Zheng et al. (2020) introduced a cloud-based co-development platform for designing smart and personalized products. In the design of these products, physical products, embedded hardware, and software with the required adaptable interface should be considered. In addition to common and configurable modules, the users could add their add-on modules. Hence, the set of possible physical, hardware, and software component parameters could be generated. A cloud-based data-driven platform makes the possibility of selecting the product attributes based on

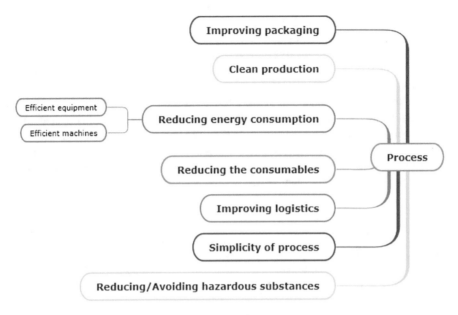

Fig. 2.12 The ecodesign practices (process perspective)

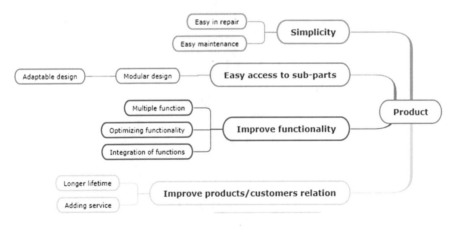

Fig. 2.13 The ecodesign practices (product perspective)—Created in Mindomo software

user's preferences and the online feedback for manufactures and designers. Zheng et al. (2019) also proposed a data-driven co-creation platform for designing the products. The user experience could be visualized in an augmented reality environment, and the manufacturer could use this real-time product feedback. Hence, the design improvement and the predictive maintenance could be derived in a cloud-based platform. In these two cases, the sustainability and eco-design perspectives could be integrated into these cloud-based and data-driven platforms. The users could test and evaluate their component parameters based on the sustainability of the product. The trade-off between costs and environmental impacts from LCA results and cost/benefits analysis could be performed to encourage the users to the green choices. These interactive co-creation platforms could increase the customers' awareness of their preferences in design, and the manufacturers could analyze the customers' behavior in design for predictive/descriptive analysis and finding the best solutions for the sustainable products. Bonvoisin (2017) discussed the open innovation 2.0 in the design of the products. This technology includes big data, IoT, and cloud computing and facilitates sharing the value between the stakeholders involved in the design of the products. The users' role is critical in the innovation of the products, and the sustainability, serviceability, and intelligence of the products could be enhanced via new business models. The applications of additive

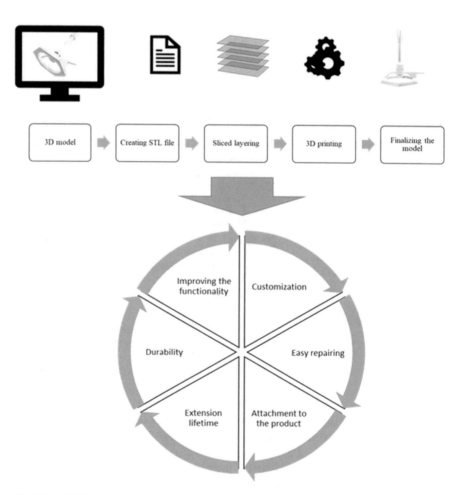

Fig. 2.14 Additive manufacturing workflow and its advantages in the product sustainability

manufacturing and 3D printing are increasing in different industrial sectors. Mami et al. (2017) discussed the eco-efficiency of 3D printing in the aerospace industry. Vanderploeg et al. (2017) studied the application of additive manufacturing in the fashion industry. There are different advantages of 3D printing including decreasing the production time, inventory costs, packaging, and warehousing. In the design stage, it accelerates the design process. Diegel et al. (2016) the role of 3D printing in sustainable design. They discussed that the customization increases the desirability of the products for the customers, and this attachment enhances the durability of the product. The topology optimization technology and new types of materials changed the design methods. According to the authors, additive manufacturing could save 90% material in the manufacturing of titanium bottle openers in comparison to the CNC machining technique. In addition, it can increase the freedom of design and flexibility and the possibility of adding any complexities at the design stage. The geometrical restrictions are not important in 3D printing, and this can increase the flexibility in manufacturing for designing the products based on the customers' preferences. The wastes during the manufacturing phase are limited and unmelted powder could be reused. It also facilitates the repair operation by easy production of spare parts. This can increase the product lifetime and improve the customer/product relation. For a broken component, a 3D model could be created by customers, and a rapid prototype could be made to test its functionality. In this case, additive manufacturing leads to extend the product lifetime and improve product features. The additive manufacturing process and its advantages in DfE are shown in Fig. 2.14. It should be noted that some limitations still exist in applying this technology such as material scarcity and costs.

2.4 Conclusion

The applications of Industry 4.0 in the eco-design context are promising. In this chapter, the eco-design trends in research and practice, the tools and methods, and the principles and the processes are discussed to highlight the future of this research field. The impacts on certain technologies of Industry 4.0 are discussed from the perspective of material, process, and product. The design stage is a critical step in the life cycle of the product. Dealing with the complexities and uncertainties of design requires collaboration between different key players at the design stage. The data and sharing platforms aid in facing these challenges. Additive manufacturing, AR/VR, and cloud-based platforms facilitate the interaction of customers and designers. Big data analytics, simulation/ optimization, and IoT enable effective and efficient decision-making in the operation and manufacturing process. Industry 4.0 and eco-design both are research fields with limited application case studies. More practical studies are required to discuss the implications of these new technologies on eco-design and address the manufacturing companies' and designers' challenges.

References

C. Bai, P. Dallasega, G. Orzes, J. Sarkis, Industry 4.0 technologies assessment: A sustainability perspective. Int. J. Prod. Econ. **229**, 107776 (2020)

T.A. Bhamra, Ecodesign: The search for new strategies in product development. Proc. Inst. Mech. Eng. B J. Eng. Manuf. **218**(5), 557–569 (2004)

T. Bhamra, R.J. Hernandez, Thirty years of design for sustainability: An evolution of research, policy, and practice. Design Sci. **7**, e2 (2021)

S.H. Bonilla, H.R. Silva, M. Terra da Silva, R. Franco Gonçalves, J.B. Sacomano, Industry 4.0 and sustainability implications: A scenario-based analysis of the impacts and challenges. Sustainability **10**(10), 3740 (2018)

J. Bonvoisin, Limits of ecodesign: The case for open source product development. Int. J. Sustain. Eng. **10**(4–5), 198–206 (2017)

M. Borchardt, L.A. Poltosi, M.A. Sellitto, G.M. Pereira, Adopting ecodesign practices: Case study of a midsized automotive supplier. Environ. Qual. Manag. **19**(1), 7–22 (2009)

M.D. Bovea, V. Pérez-Belis, A taxonomy of ecodesign tools for integrating environmental requirements into the product design process. J. Clean. Prod. **20**(1), 61–71 (2012)

E. Cagno, A. Neri, M. Negri, C.A. Bassani, T. Lampertico, The role of digital technologies in operationalizing the circular economy transition: A systematic literature review. Appl. Sci. **11**(8), 3328 (2021)

P. Cicconi, Eco-design and eco-materials: An interactive and collaborative approach. Sustain. Mater. Technol. **23**, e00135 (2020)

B.S. Costa, A.C. Bernardes, J.V. Pereira, V.H. Zampa, V.A. Pereira, G.F. Matos, ..., A.F. Silva, Artificial intelligence in automated sorting in trash recycling, in *Anais do XV Encontro Nacional de Inteligência Artificial e Computacional* (SBC, 2018), pp. 198–205.

S. da Costa Fernandes, D.C. Pigosso, T.C. McAloone, H. Rozenfeld, Towards product-service system oriented to circular economy: A systematic review of value proposition design approaches. J. Clean. Prod. **257**, 120507 (2020)

J. de Aguiar, L. de Oliveira, J.O. da Silva, D. Bond, R.K. Scalice, D. Becker, A design tool to diagnose product recyclability during product design phase. J. Clean. Prod. **141**, 219–229 (2017)

O. Diegel, P. Kristav, D. Motte, B. Kianian, Additive manufacturing and its effect on sustainable design, in *Handbook of sustainability in additive manufacturing*, (Springer, Singapore, 2016), pp. 73–99

E. Dostatni, J. Diakun, D. Grajewski, R. Wichniarek, A. Karwasz, Automation of the ecodesign process for Industry 4.0, in *International Conference on Intelligent Systems in Production Engineering and Maintenance*, (Springer, Cham, 2018a), pp. 533–542

E. Dostatni, I. Rojek, A. Hamrol, The use of machine learning method in concurrent ecodesign of products and technological processes, in *Advances in Manufacturing*, (Springer, Cham, 2018b), pp. 321–330

Ellen MacArthur Foundation, Artificial Intelligence and the Circular Economy: AI as a Tool to Accelerate the Transition, 2019. https://www.ellenmacarthurfoundation.org/assets/downloads/Artificial-intelligence-and-the-circular-economy.pdf

C. Enyoghasi, F. Badurdeen, Industry 4.0 for sustainable manufacturing: Opportunities at the product, process, and system levels. Resour. Conserv. Recycl. **166**, 105362 (2021)

J. Faludi, S. Hoffenson, S.Y. Kwok, M. Saidani, S.I. Hallstedt, C. Telenko, V. Martinez, A research roadmap for sustainable design methods and tools. Sustainability **12**(19), 8174 (2020)

Y. Feng, Y. Zhao, H. Zheng, Z. Li, J. Tan, Data-driven product design toward intelligent manufacturing: A review. Int. J. Adv. Robot. Syst. **17**(2), 1729881420911257 (2020)

M. Ghobakhloo, Industry 4.0, digitization, and opportunities for sustainability. J. Clean. Prod. **252**, 119869 (2020)

F. Gu, B. Ma, J. Guo, P.A. Summers, P. Hall, Internet of things and Big Data as potential solutions to the problems in waste electrical and electronic equipment management: An exploratory study. Waste Manag. **68**, 434–448 (2017)

M. Holland, J. Stjepandić, C. Nigischer, Intellectual property protection of 3D print supply chain with blockchain technology, in *2018 IEEE International Conference on Engineering, Technology and Innovation (ICE/ITMC)* (IEEE, 2018), pp. 1–8

S.S. Kamble, A. Gunasekaran, S.A. Gawankar, Sustainable Industry 4.0 framework: A systematic literature review identifying the current trends and future perspectives. Process Saf. Environ. Prot. **117**, 408–425 (2018)

G. Kane, *Building a sustainable supply chain* (Routledge, 2017)

A. Karwasz, J. Trojanowska, Using CAD 3D system in ecodesign—case study, in *Efficiency in Sustainable Supply Chain*, (Springer, Cham, 2017), pp. 137–160

S. Keivanpour, D.A. Kadi, Perspectives for application of the internet of things and big data analytics on end of life aircraft treatment. Int. J. Sustain. Aviat. **4**(3–4), 202–220 (2018)

H. Kim, F. Cluzel, Y. Leroy, B. Yannou, G. Yannou-Le Bris, Research perspectives in ecodesign. Design Sci. **6**, e7 (2020)

J. Kim, M. Saidani, H.M. Kim, Designing an optimal modular-based product family under intellectual property and sustainability considerations. J. Mech. Design **143**(11), 112002 (2021)

V. Lofthouse, Ecodesign tools for designers: Defining the requirements. J. Clean. Prod. **14**(15–16), 1386–1395 (2006)

C. Luttropp, J. Lagerstedt, EcoDesign and the ten golden rules: Generic advice for merging environmental aspects into product development. J. Clean. Prod. **14**(15–16), 1396–1408 (2006)

F. Mami, J.P. Revéret, S. Fallaha, M. Margni, Evaluating eco-efficiency of 3D printing in the aeronautic industry. J. Ind. Ecol. **21**(S1), S37–S48 (2017)

F. Mathieux, D. Brissaud, P. Zwolinski, Product ecodesign and materials: Current status and future prospects. arXiv preprint arXiv:0711.1788 (2007)

Mistbreaker Website, (n.d.), http://www.mistbreaker.com/sustainability/artificial-intelligence-put-use-recycling/

J. Oláh, N. Aburumman, J. Popp, M.A. Khan, H. Haddad, N. Kitukutha, Impact of Industry 4.0 on environmental sustainability. Sustainability **12**(11), 4674 (2020)

D.C.A. Pigosso, H. Rozenfeld, T.C. McAloone, Ecodesign maturity model: A management framework to support ecodesign implementation into manufacturing companies. J. Clean. Prod. **59**, 160–173 (2013)

D.C.A. Pigosso, T.C. McAloone, H. Rozenfeld, Characterization of the state-of-the-art and identification of main trends for Ecodesign tools and methods: Classifying three decades of research and implementation. J. Indian Inst. Sci. **95**(4), 405–428 (2015)

S. S. Rattan, Strength of materials, Tata McGraw-Hill Education, 2008, https://books.google.ca/books/about/Strength_of_Materials.html?id=oXBLj7Jrr7YC

I. Rojek, E. Dostatni, Artificial neural network-supported selection of materials in ecodesign, in *International Scientific-Technical Conference Manufacturing*, (Springer, Cham, 2019), pp. 422–431

I. Rojek, E. Dostatni, Machine learning methods for optimal compatibility of materials in ecodesign. Bull. Polish Acad. Sci. Techn. Sci., **68**(2) (2020)

I. Rojek, D. Mikołajewski, E. Dostatni, Digital twins in product lifecycle for sustainability in manufacturing and maintenance. Appl. Sci. **11**(1), 31 (2021)

M. Rossi, M. Germani, A. Zamagni, Review of ecodesign methods and tools. Barriers and strategies for an effective implementation in industrial companies. J. Clean. Prod. **129**, 361–373 (2016)

R.K. Scalice, D. Becker, R.C. Silveira, Developing a new compatibility table for design for recycling. Product: Management and Development **7**(2), 141–148 (2009)

K. Schischke, M. Hagelüken, S. Bai, fenhagen, G., An introduction to ecodesign strategies–why, what and how?. Fraunhofer IZM, Berlin, Germany. Ecodesign implementation into manufacturing companies. J. Clean. Prod. **59**(2013), 160–173 (2005)

C. Thomas, I. Maître, R. Symoneaux, Consumer-led eco-development of food products: A case study to propose a framework. Br. Food J. (2021)

J. Trojanowska, A. Karwasz, J.M. Machado, M.L.R. Varela, Virtual reality based ecodesign, in *Efficiency in Sustainable Supply Chain*, (Springer, Cham, 2017), pp. 119–135

C. Van Hemel, J. Cramer, Barriers and stimuli for ecodesign in SMEs. J. Clean. Prod. **10**(5), 439–453 (2002)

A. Vanderploeg, S.E. Lee, M. Mamp, The application of 3D printing technology in the fashion industry. Int. J. Fashion Design Technol. Educ. **10**(2), 170–179 (2017)

Y. Zhang, S. Ren, Y. Liu, S. Si, A big data analytics architecture for cleaner manufacturing and maintenance processes of complex products. J. Clean. Prod. **142**, 626–641 (2017)

P. Zheng, Y. Lin, C.H. Chen, X. Xu, Smart, connected open architecture product: An IT-driven co-creation paradigm with lifecycle personalization concerns. Int. J. Prod. Res. **57**(8), 2571–2584 (2019)

P. Zheng, X. Xu, C.H. Chen, A data-driven cyber-physical approach for personalised smart, connected product co-development in a cloud-based environment. J. Intell. Manuf. **31**(1), 3–18 (2020)

Chapter 3
Design for End of Life 4.0

3.1 Introduction

Population growth, the variety of the products, and technology lead to more consumption of materials and natural resources and products with a shorter life cycle.
The products reach to EoL phase, and the management of these products plays an
essential role in sustainability and circular economy. Reducing, reusing and recycling are the basis for different EoL strategies. The recovery logistics of the EoL
products is a complex research problem considering the uncertainty in the flow of
the material, different players who are involved in the recovery network, and the
data availability. Some of these challenges could be managed at the design stage
with sustainable material selection, improving the architecture of the products for
improving the disassembly and material recovery and focusing on material recyclability. With more customer's attachment to the products and involving them at
the design stage, we could expect the products with longer lifetime. We should create the connection of the customers to the products during the whole life cycle. In
this chapter, we review the literature of design for EoL and the new trends in EoL
practices. Then, the implications of Industry 4.0 are discussed. The applications via
a codesign of an airplane toy are discussed to highlight the role of data analytics,
cloud, and AR in enhancing the sustainable design for the circular economy.

3.2 Literature Review

Design for EoL is one of the most important DfX practices in the context of the
circular economy. Different EoL solutions should be considered and analyzed at the
design stage of the products. Figure 3.1 summarizes these solutions and their characteristics (Atlason et al. 2017). Figure 3.2 shows the word cloud of some parts of

Reusing	Refurbishing	Remanufacturing	Recycling	Energy recovery	Disposal
• Little changes or modifications • Using the products for the same purpose	• Performing cleaning, repairing, painting or other modifications for improving the quality of the used products for reuse	• Modification and recovery at subassembly level • Certifications and quality assurance are required	• Destructive disassembly operation • Recovery of the materials	• Produced energy from EoL parts and material • The energy level depend on heat values from each type of parts or materials	• Less preferable option • The related costs of safe disposal should be considered

Fig. 3.1 Different EoL strategies

Fig. 3.2 Word cloud (some parts of reviewed articles in this book including abstract, introduction, and conclusion)—created in https://worditout.com/word-cloud/create

reviewed articles in this book including abstract, introduction, and conclusion. As shown in the figure, data analysis, EoL, process, materials, disassembly are the main keywords in the research. This could highlight the role of the circular economy in DfE literature and develop the methods and tools in this context.

In this section, a review of the studies on design for EoL is provided. The aims of this review are as follows:

- Highlighting the gaps and the trends in design for EoL
- Survey on recent research on eco-design in the context of circular economy
- Study on the applications of Industry 4.0 in the context of design for EoL

3.2.1 Design for EoL

Favi et al. (2017) proposed a framework for design for EoL. The framework includes three main steps:

- Step 1 includes disassembling analysis including 3D model analysis and BoM for developing optimized disassembling scenarios.
- Step 2 includes EoL indices calculation for different EoL options (reusing, remanufacturing, recycling, and disposal).
- Step 3 is redesigning the products based on the results of step 2. In this phase, based on EoL indices, the minor redesign or deep redesign will be performed. The minor redesign includes modification in structure and connection, while the deep redesign includes the change in material, change in product architecture, shape, or geometry of the product.

Anthony and Cheung (2017) explained that the disassembly and EoL strategies are becoming more important in the automotive industry. The reasons are including recovery of the valuable parts, removing the hazardous material, and separation processes for the valorization of the material. The authors analyzed the costs of different EoL options: remanufacturing, refurbishing, recycling, and disposal. Different strategies can be considered at the design stage for reducing the EoL treatment costs as follows:

- Facilitating and expedition of dissembling process with reducing the number of assemblies
- Reducing the time of disassembly with simple fasteners or reducing fastening
- Avoiding mixed material for improving the material recovery
- Automating the disassembling process

Cheung et al. (2015) proposed a framework for predicting the EoL costs of the product at the design stage. They used a cube concept for addressing the cost elements. The cost element should be determined based on cost categories, BoM, and the life cycle of the product. The cost categories are disassembly, cleaning, testing, reprocessing, assembling, and final testing.

Ma et al. (2018) discussed the importance of dealing with uncertainties in the decision-making of EoL at the design stage. They explained the EoL strategies could be varied based on geographic locations and the users' preferences. They proposed a fuzzy decision-making approach for considering the uncertainties of EoL options.

Mandolini et al. (2019) emphasized the role of the data model in facilitating the eco-design of complex products.

3.2.2 DfE and Circular Economy

Atlason et al. (2017) discussed the product design strategies in the context of the circular economy. The extended circularity could be possible via reusing the products, parts, and material in different forms after their first lifetime. For design for EoL, the role and preferences of the users are essential. The strategies should be aligned based on prioritizing the customers' choice concerning the EoL products. The author explained, for example, when the customers prefer repairing options, the effort for recyclability of the product at the design stage is not a successful EoL strategy. Improving the guideline for repairing and facilitating the accessibility to spare parts are more appropriate strategies in this case. They suggested adopting the EoL strategies based on the type of the products and users' characteristics (demographic, attitude, willingness to pay, risk-taking, etc.). The authors used the Kano model to analyze the role of users in selecting the best EoL strategies. In this model, the satisfaction of the users is a function of the product's features. The lack of "must-be" features leads to dissatisfaction of the users. For example, installed windows and antivirus applications are mandatory features for a new laptop. Improving one-dimensional features can increase user satisfaction. The warranty period of the laptop or battery life is a one-dimensional feature of the product. The attractive feature is the product's features that are not expected by users, and adding these features could lead to increased user satisfaction. The installed Microsoft package or any accessories for the laptop can increase user satisfaction. Hence, the EoL strategies should be aligned with users' characteristics and their satisfaction based on the different features of the products (see Fig. 3.3).

Bauer et al. (2021) discussed the design for cascading and the new trends in design for EoL. Hence, several cascaded value chains for the products could be considered. The products could be used for different purposes and applications. The design process includes analyzing the customers' requirements, developing the product specifications, conceptual and detailed design, testing, and manufacturing. After the first lifetime, another design process could be planned. In the second lifetime, again the customers' requirements should be analyzed and the conceptual and detailed design, manufacturing, and logistics related to the second life of the products should be developed. The elements of Porter's value chain (Porter 2001) including primary and secondary activities are shown in Fig. 3.4. For example, for a facial mask, the ear straps could be used for the hair ties as the second use, and even the third life could be designed (reusing as USB cable rap or using the material in a cushion; see Fig. 3.5).

Den Hollander et al. (2017) also discussed product design in the context of the circular economy. They defined circular design as design for product integrity and

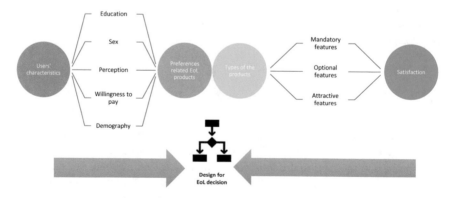

Fig. 3.3 The role of products users in design for EoL decision

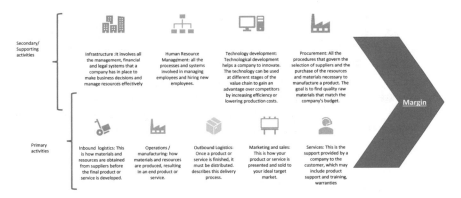

Fig. 3.4 The Porter's value chain

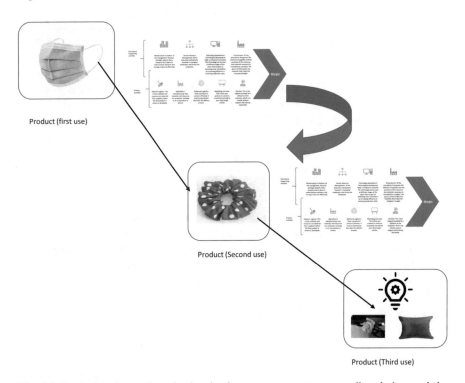

Product (first use)

Product (Second use)

Product (Third use)

Fig. 3.5 Designing the products in the circular economy context, cascading design, and the value chain

design for recycling. Design for product integrity divides into three types of strategies: design for longer use, design for extended use, and design for recovery.

Design for longer use includes the efforts or increasing the physical durability of the products and the emotional attachment of the products to users with personalization. Design for extended use includes design for maintenance and upgrading.

Design for recovery includes the design strategies for repairing, refurbishing, or remanufacturing.

Favi (2021) discussed the product design in a circular economy via the case study of the coffee machine. The author mentioned that few business models discussed the circular eco-design. Disassembly is more discussed in comparison to the business models for recycling and remanufacturing. He also emphasized that developing a performance evaluation framework for circular eco-design with integrating three aspects of sustainability is a research gap in the existing literature.

3.2.3   Industry 4.0 and Design for EoL

Cagno et al. (2021) discussed the role of digital technologies in the circular economy. The authors used a six elements framework of ReSOLVE (Revegetate, Share, Optimize, Loop, Virtualize, and Exchange) for addressing the implications of Industry 4.0 in the circular economy. IoT positively influences the product lifetime by monitoring and controlling the product during its life cycle. IoT also with the opportunity for tracing and tracking supports reusing and recycling of the parts and materials. Bigdata analytics facilitates reverse logistics operation via sharing information among the value chain actors and facilitating real-time decision-making. Digital technologies and autonomous robots facilitate the disassembly processes and sorting operation. Simulation and big data analytics also aid in data mining and improving disassembly sequence planning.

Jiao et al. (2021) discussed design engineering in the era of Industry 4.0. The authors discussed that the fourth industrial revolution leads to designing more personalized products, smart and connected products, data-driven decision-making, and co-creation of the products. They mentioned that the future trend on circular eco-design should be focused on design at the upstream and product use at the downstream level. Eco-design of the products should be done with involving EoL stakeholders and data-driven approaches for analyzing recyclability.

Joshi and Gupta (2019) proposed an optimization model for evaluating the design for EoL using IoT. They used the traceability of IoT as a mechanism in the evaluation of the feasibility of recovery options. The authors used MCDM and linear programming to evaluate the design options for ease of disassembly based on the total profit, quality level, and the number of disposed items.

The students of the Eindhoven University of Technology with the collaboration of Oceanz designed Noah with 3D printing technology. In this passenger car, the interior design based on ecofriendly materials and facilitating disassembly operation are considered (Noah n.d.). There are several examples of using bio-based material in the interior design of the passenger car. Porsche's Taycan interior is designed based on a driver-focused approach, and olive leaves are used for the sustainability of tanned club leather (Porsche n.d.). Hyundai also used bioplastic in the interior design of the Kia Soul EV. It is also used in the Nexo Fuel Cell dashboard (Hyundai n.d.) (See Fig. 3.6). One essential element in designing in a customer-centric paradigm is

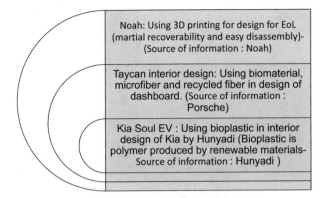

Fig. 3.6 Examples of Design for EoL in the automotive industry

collaborative design. The interior design of the new cars should be based on a better user experience, and the interactive platform plays an essential role in integrating the customers' preferences. In the customized interior design of passenger cars, the integration of sustainability via a collaborative platform between designers, manufacturers, and the customers could improve the users' experience. The recyclability and recoverability of the cars are important factors in the design for EoL. According to Torn and Vaneker (2019), personalization adds different challenges to the manufacturing process and supply chain. One of these challenges is communicating real-time information with all stakeholders during codesign process. Industry 4.0 with vertical and horizontal integration, big data analytics, and cloud computing facilitates a sustainable design process.

3.3   Design for EoL 4.0: A Conceptual Framework

Based on the synthesis of the literature review, three elements of (1) the circular value chain, (2) optimizing design for EoL practices, and (3) Industry 4.0 could create a synergy for circular design in the future (see Fig. 3.7). Designing the cascade value chain that includes the interconnection of the value between the first and the next life of the products is essential in future circular business models. The role of stakeholders, their involvement in collaborative design, and designing appropriate value networks is vital. One of the key stakeholders is customers. Their references, characteristics, and contributions facilitate the sustainability of design for the EoL chain. The literature of design for EoL has enriched studies on tools and methods for the extension of product lifetime and design strategies for optimizing the valorization of material and parts. The trade-off between costs and environmental impacts are frequently considered in the literature. However, the uncertainties in the reverse flow in terms of quality, quantity, and the other conditions of EoL products should be integrated into the decision tools. Industry 4.0 technologies could boost the

Fig. 3.7 The proposed
conceptual model

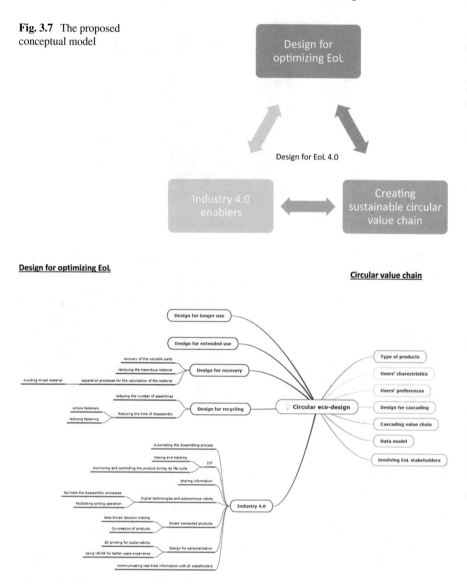

Fig. 3.8 The key elements of the conceptual model

sustainability of the value chain and addressing the uncertainties of the returned
flows with transparency of the life cycle and intelligent decision support tools. IoT
aids in monitoring and controlling the products during the life cycle. VR and AR
and 3D printing facilitate the integration of customers in the codesign and co-
creation process. Big data analytics aids in dealing with uncertainties of the cascaded value chain and intelligent systems accelerate the decision-making process.

Digital technologies and autonomous robots help in the automation of the dissembling process and sorting operation and the management of the recovery channels. The valuable feedback from these operations aids in optimizing the design stage. Figure 3.8 shows the elements of the conceptual framework.

3.4 Application Perspective: An Illustrative Example

Gamification has received much attention in the circular economy context and for changing the users' behavior. It can use as motivating leverage for sustainability activities. Several studies recently explained the application of gamification for recycling wastes and encouraging adults to separate wastes stream (e.g., Briones et al. 2018; Aguiar-Castillo et al. 2019). In the 3R approach, reducing is the highest value activity for social responsibility. Hence, eco-design and design for EoL are the preferable practices that could be enhanced and encouraged via gamification. In this section, a hypothetical example of designing an airplane toy is illustrated to highlight the application of Industry 4.0 technologies in boosting sustainable codesign activities. The life cycle analysis of toys is not well studied in the literature. Recycling electronic parts and WEEE batteries are studied in the literature (Muñoz et al. 2009). Muñoz et al. (2009) analyzed the LCA of a teddy bear toy and recommended eco-design solutions based on the hotspot results. The author explained that the use phase considering the types of batteries has the highest impact. The first step in integrating the environmental consideration in product design is considering the life cycle of the target product. The main elements in BoM of airplane toys are shown in Fig. 3.9. Hence, the material and parts production, the assembly, packaging, shipping, and end of life are the most important phases in the LCA of this toy (see Fig. 3.10 for LCA scope). The electric/electronic and batteries could also be added for an electronic toy option. Codesign and co-creative design is a new paradigm in the design of consumable products with considering the active role of end

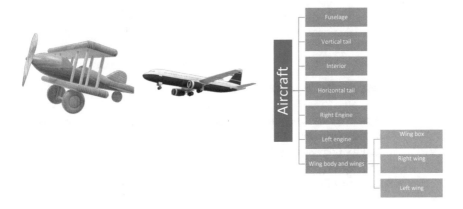

Fig. 3.9 Airplane toy and BoM

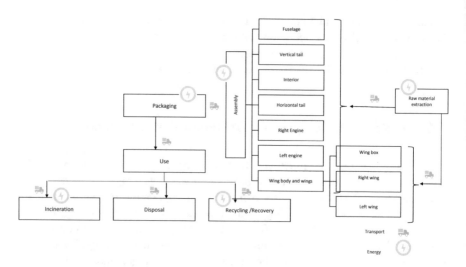

Fig. 3.10 The main elements in the life cycle of an airplane toy

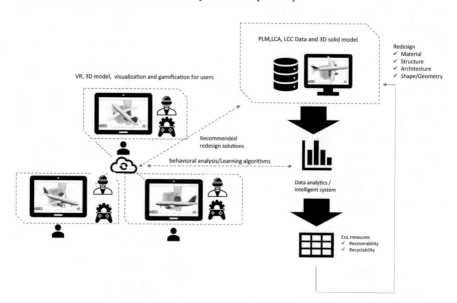

Fig. 3.11 The proposed architecture for codesign of an airplane toy

users. Augmented reality and virtual reality are widely used in this context to enhance the user experiments in the designing stage. Cascini et al. (2020) developed a VR-based tool for codesign of packaging. Yun and Leng (2021) also used VR and CAD tools for designing product packaging.

Figure 3.11 shows the building blocks of codesign architecture. The users have access to an interactive dashboard for visualization, 3D modeling, VR/AR experiment, and gamification.

The designed prototypes are saved on the cloud and will be accessible by designers to evaluate them based on material, structure, architecture, and shape/geometry. The designers' decision tools have access to different databases including PLM, LCA, LCC, and 3D solid model. Several models could be tested via a data analytical module for calculating the recyclability and recoverability of the model. The performance of the designed model will be communicated to users and designers for the required design adjustment. The data analytics module also uses the learning algorithm for analyzing the patterns of design based on the user's characteristics. This information helps manufacturers for predicting the users' demand in the future and analyze the users' preferences based on AI.

The user dashboard architecture is shown in Fig. 3.12. Users could select each part of the toy and then change the dimension, shape, weight, color, and material. For example, for materials, they can select different types of plastic, metal, wood, or fabric. An interactive dashboard is designed for users to see the final product 3D model and the product performance in terms of environmental impacts, life cycle costs, and safety score. Then, a score will be assigned to the user based on this design. The user could save this design and compare it with the other design options for improving the score and the performance of the designed model.

For the data analytics part, several approaches including decision tree, fuzzy inference system, and clustering can be used for the evaluation of the DfE score of the designed model. Here, we explain an example of score evaluation based on six input variables via a fuzzy simulation model. A Mamdani FIS (fuzzy inference system) is developed in MATLAB environment. The DfE score is a function of

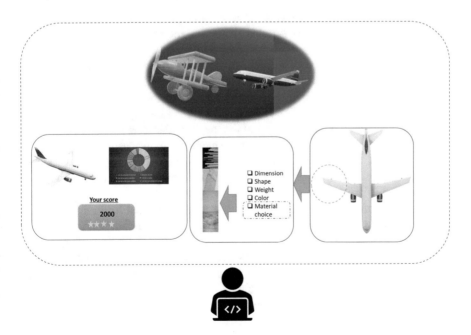

Fig. 3.12 User's dashboard

material sustainability, easiness in disassembly, recyclability, complexity at suppliers, and manufacturing levels and costs (Eq. 3.1).

$$DfE_{Score} = FIS(\widetilde{M.S}, \widetilde{diss}, \widetilde{Recy}, \widetilde{c_{sup}}, \widetilde{c_{man}}, \widetilde{cost})$$

(3.1)

The variables are as follows:

Variables	Description	Range
$\widetilde{DfE\_Score}$	*The final score of DfE*	*(Green, Yellow, Red)*
$\widetilde{M.S}$	*Material sustainability*	*(Low, Medium, High)*
$\widetilde{diss}$	*Easiness in disassembly*	*(Low, Medium, High)*
$\widetilde{Recy}$	*Recyclability*	*(Low, Medium, High)*
$\widetilde{c_{sup}}$	*Complexity at suppliers level*	*(Low, Medium, High)*
$\widetilde{c_{man}}$	*Complexity at the manufacturing level*	*(Low, Medium, High)*
$\widetilde{cost}$	*Total LCC costs score*	*(Low, Medium, High)*

The triangle fuzzy membership function is considered for each input variable. The crisp input variable of the FIS model should be transferred into a linguistic value. The defuzzification process will be performed to build a numeric value. The fuzzy rules should be defined based on the sustainability objective of manufactures. The interpretation of the rules depends on the designers' teams based on the enterprise priorities. It determines the level of trade-off analysis between the input variables. The generated rules depend on the number of input variables in evaluating the DfE score and the membership functions for each input. The FIS model with six input variables is shown in Fig. 3.13. The examples of the fuzzy rules and the surface of the rules are shown in Figs. 3.14 and 3.15, respectively. The (if-part) of each rule is selected based on the priorities of the toy manufacturers and the design team. The (then-part) will be assigned based on the judgment of the designers/expert team.

For example, if the material sustainability is green, the easiness of disassembly is high, the recyclability is high, the complexity for suppliers and manufacturing is low and the LCC is low, the DfE score will be green.

The user scores in the gamification platform are calculated based on the design history and the DfE score of the suggested design. The users based on the number of (Green, Yellow, Red) designs will get points in the system. A rewarding mechanism could be designed based on the total score of the users. The rewarding mechanism could be defined as a gift card or price reduction in the planned purchase or future purchasing. It should be noted that this proposed platform could also be integrated into online retailers' cloud-based platforms. AR capability enhances the users' experience in designing the target toy. A selected 3D model file could be

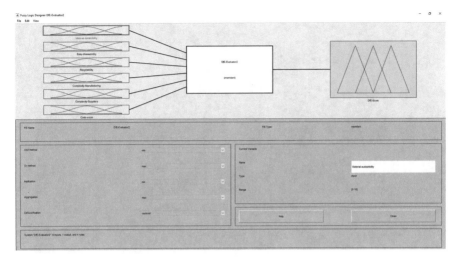

Fig. 3.13 FIS model for evaluation of DfE score of the proposed design

Fig. 3.14 The examples of generated rules

entered into 3D modeling software, and using augmented reality software, 3D content could be created. With an AR camera and real-time 3D platform, testing the model in real time will be possible.

The real-time testing in the 3D platform allows the users to see different types of packaging and design options and customize the product based on their preferences.

Hence, the proposed platform provides the following benefits:

- Participating the users in design actively
- Enhancing the users' experience

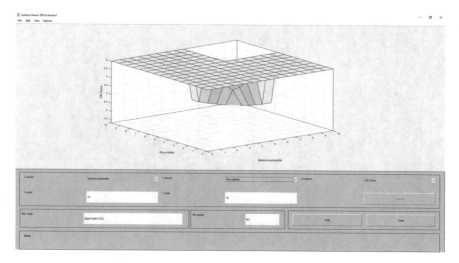

Fig. 3.15 The surface of the rules

- Giving the feedback of sustainability of the design options
- Motivating users for the sustainability choices
- Collaborating between key players
- Integrating different databases for multidimensional analysis
- Using data analytics for visualization and processing a large amount of data

3.5 Conclusion

This chapter discussed the design for EoL in Industry 4.0 and the customer-centric paradigm. The different approaches for EoL strategies at the design stage are explained. The perspective of applying different Industry 4.0 technologies is discussed. A hypothetical example in co-creative design for an airplane toy is provided to propose a conceptual framework for DfE.4.0 and considering the circular economy aspect. There are some challenges in the applications of the proposed approaches that should be addressed in future research. For example, designing an interactive decision dashboard for users and data exchange between designers and users require integrating different databases and sometimes lead to modeling complexity. Using simulation and developing metamodels could aid in this context. The application in a real case study or an example for open-architecture products is proposed as future research.

References

L. Aguiar-Castillo, A. Clavijo-Rodriguez, D. Saa-Perez, R. Perez-Jimenez, Gamification as an approach to promote tourist recycling behavior. Sustainability **11**(8), 2201 (2019)

C. Anthony, W.M. Cheung, Cost evaluation in design for end-of-life of automotive components. J. Remanufac. **7**(1), 97–111 (2017)

R.S. Atlason, D. Giacalone, K. Parajuly, Product design in the circular economy: Users' perception of end-of-life scenarios for electrical and electronic appliances. J. Clean. Prod. **168**, 1059–1069 (2017)

T. Bauer, G. Mandil, É. Monnier, P. Zwolinski, Design for cascading applications reuse–understandings of an emerging end-of-use strategy and propositions for its implementation. J. Eng. Des. **32**, 140–163 (2021)

A.G. Briones, P. Chamoso, A. Rivas, S. Rodríguez, F. De La Prieta, J. Prieto, J.M. Corchado, Use of gamification techniques to encourage garbage recycling. A smart city approach, in *International Conference on Knowledge Management in Organizations*, (Springer, Cham, 2018), pp. 674–685

E. Cagno, A. Neri, M. Negri, C.A. Bassani, T. Lampertico, The role of digital technologies in operationalizing the circular economy transition: A systematic literature review. Appl. Sci. **11**(8), 3328 (2021)

G. Cascini, J. O'hare, E. Dekoninck, N. Becattini, J.F. Boujut, F.B. Guefrache, et al., Exploring the use of AR technology for co-creative product and packaging design. Comput. Ind. **123**, 103308 (2020)

W.M. Cheung, R. Marsh, P.W. Griffin, L.B. Newnes, A.R. Mileham, J.D. Lanham, Towards cleaner production: A roadmap for predicting product end-of-life costs at early design concept. J. Clean. Prod. **87**, 431–441 (2015)

M.C. Den Hollander, C.A. Bakker, E.J. Hultink, Product design in a circular economy: Development of a typology of key concepts and terms. J. Ind. Ecol. **21**(3), 517–525 (2017)

C. Favi, Product eco-design in the era of circular economy: Experiences in the design of Espresso coffee machines, in *Advances on Mechanics, Design Engineering and Manufacturing III: Proceedings of the International Joint Conference on Mechanics, Design Engineering & Advanced Manufacturing, JCM 2020, June 2–4, 2020* (Springer Nature, 2021), p. 194

C. Favi, M. Germani, A. Luzi, M. Mandolini, M. Marconi, A design for EoL approach and metrics to favour closed-loop scenarios for products. Int. J. Sustain. Eng. **10**(3), 136–146 (2017)

Hyundai, (n.d.). https://tech.hyundaimotorgroup.com/article/eco-friendly-cars-to-become-actually-eco-friendly-materials/

J.R. Jiao, S. Communri, J.H. Panchal, J. Milisavljevic-Syed, J.K. Allen, F. Mistree, D. Schaefer, Design engineering in the age of Industry 4.0. J. Mech. Des. **142**(8), 088001 (2021)

A.D. Joshi, S.M. Gupta, Evaluation of design alternatives of end-of-life products using internet of things. Int. J. Prod. Econ. **208**, 281–293 (2019)

J. Ma, G.E.O. Kremer, C.D. Ray, A comprehensive end-of-life strategy decision making approach to handle uncertainty in the product design stage. Res. Eng. Des. **29**(3), 469–487 (2018)

M. Mandolini, M. Marconi, M. Rossi, C. Favi, M. Germani, A standard data model for life cycle analysis of industrial products: A support for eco-design initiatives. Comput. Ind. **109**, 31–44 (2019)

I. Muñoz, C. Gazulla, A. Bala, R. Puig, P. Fullana, LCA and ecodesign in the toy industry: Case study of a teddy bear incorporating electric and electronic components. Int. J. Life Cycle Assess. **14**(1), 64–72 (2009)

Noah, (n.d.). https://blog.drupa.com/en/the-future-of-sustainable-automotive-design-2/

Porsche, (n.d.). https://www.globenewswire.com/news-release/2019/08/22/1905796/0/en/Digital-clear-sustainable-the-interior-of-the-new-Porsche-Taycan.html

M.E. Porter, The value chain and competitive advantage. Understand. Bus. Proc. **2**, 50–66 (2001)

I.A.R. Torn, T.H. Vaneker, Mass personalization with Industry 4.0 by SMEs: A concept for collaborative networks. Procedia Manufac. **28**, 135–141 (2019)

Yun, Q., & Leng, C. (2021). *Using VR Technology Combined with CAD Software Development to Optimize Packaging Design*

Chapter 4
Lean and Industry 4.0 Implication in Circular Design: An Operational Performance Perspective

4.1 Introduction

Operational excellence aids manufacturers in making good collaboration with the other players in the supply chain, create value for customers and improve the position of the companies in the dynamic and competitive market. Operational performance is a key issue for the adoption of new technologies and management practices. The lean and green supply chain have received much attention in the literature review due to the essential role of lean management in the three pillars of sustainability. Industry 4.0 as a new industrial revolution also influenced the product life cycle, and research on the implications of different digital technologies on manufacturing is growing. Lean, sustainability, and Industry 4.0 create a triangle of operational excellence, and the number of studies that consider this interaction is increasing. However, the study on analyzing this triangle from a life cycle cost perspective is limited. This chapter aims to take a step toward assessing the interaction among circular economy, lean management, and Industry 4.0 via the life cycle cost approach. The probabilistic total cost of ownership is used to provide a simulation for measuring operational performance. The contribution of this study is summarized as follows:

- Using a probabilistic total cost of ownership (TCO) model for addressing the operational performance of lean, Industry 4.0, and circular economy
- Developing a Monte Carlo simulation for analyzing the life cycle cost with integrating eco-design and product recovery elements

The rest of the chapter is organized as follows: Sect. 4.2 provides a brief review of studies that focused on the link between lean, Industry 4.0, and circular economy.

Section 4.3 provides the proposed approach, Sect. 4.4 illustrates a numerical example, and finally, Sect. 4.4 concludes with some remarks and future research direction.

4.2   Literature Review

Indeed, in a constantly changing world, you have to be able to react as quickly as possible, to adapt, to progress. The goal is no longer to manage "change" but to manage "movement". Change often involves moving from one stable state to another. But it is not a stable state that one should seek, rather it is a positive imbalance that keeps the company moving forward. The concept of Lean is based on the search for products that perfectly meet customer expectations, at exceptionally low costs and of exceptional quality. (Courtois et al. 2011: 305)

The primary representation of lean was based on the Toyota Production System known as "house of lean" with two pillars including Just in Time and Jidoka. The foundation includes Kaizen, Heijunka, standardization of tasks and stability, and the overall objective for production with minimum costs, delay, and high quality. The actual model of lean elements and the principle outcomes are shown in Fig. 4.1. Different aspects of lean management can be discussed from the perspective of sustainability and circular design. These impacts are addressed based on three pillars of sustainability in Fig. 4.2.

In this section, a brief review of some studies that focused on the relationship of Industry 4.0, lean management, and sustainability is provided. For this purpose, the combinations of the following keywords are used:

- The impacts of lean on sustainability
- The impacts of Industry 4.0 on sustainability
- The impacts of lean on Industry 4.0
- The links among lean, Industry 4.0, and sustainability

It should be noted that close keywords like circular economy, green manufacturing, sustainable operations are also considered. Most of the papers are published between 2018 and 2021. So, it shows that this area of research is at the infancy stage and needs more theoretical and practical contributions. Cagnetti et al. (2021)

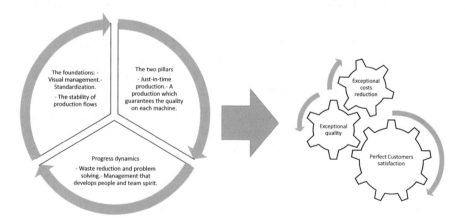

Fig. 4.1 Lean principles and the outcomes

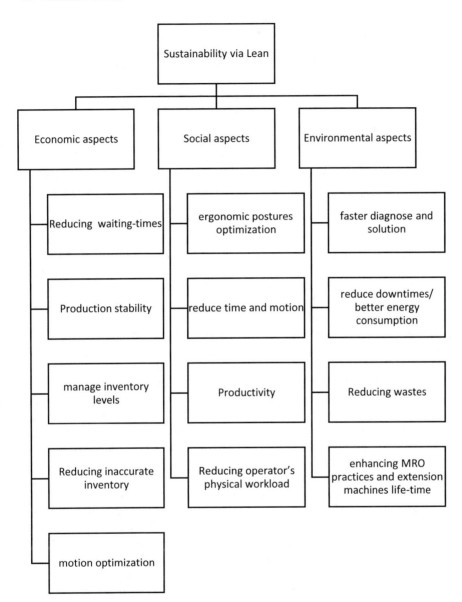

Fig. 4.2 Sustainability through lean production

addressed the link between lean production and Industry 4.0 from strategy and technical perspectives. The authors mentioned the core benefits of lean management as improving productivity, reducing operation costs, improving social sustainability, and reducing wastes and environmental impacts. Industry 4.0 could enhance information sharing and leads to better productivity and flexibility. The weakness of lean management including demand fluctuation and poor customization also could be improved by Industry 4.0 technologies. Dev et al. (2020) studied the implications of Industry 4.0 on circular economy for operational excellence. The authors focused on remanufacturing operation and the impacts of Industry 4.0 for improving the scheduling, lot size, setup time, and order release. They used virtual cellular manufacturing and clustering the products based on setup time requirements. The application of additive manufacturing could enhance the remanufacturing operation performance. Durakovic et al. (2018) performed a literature review on lean management and focused on trends, benefits, challenges, and opportunities. The authors concluded that lean is well established and used in automotive, aerospace, metal, and textile industries as well as hospitals and ergonomic contexts. The benefits include reducing wastes, production time, improving financial performance, and sustainability. Some challenges are related to monitoring, management commitment, quality control, and inventory management that could be improved by transparency, information sharing, and optimization as the result of digital transformation. The easy integration of lean with the new methods and procedures are the advantages that could be considered as the opportunities of this approach. Kurdve and Bellgran (2021) studied the implications of green and lean on the circular economy. The outcomes are classified as increasing resource utilization, using substitute material and energy, avoiding non-value-added consumption, and contributing to a closed-loop material chain. Pagliosa et al. (2019) assessed the link between lean and Industry 4.0. They mentioned that categorization of different technologies in Industry 4.0 and lean management in different levels of the value stream is required. They also explained that few studies addressed the impact of Industry 4.0 and lean management on operational performance. Piercy and Rich (2015) evaluated the link between lean operation and sustainability. The authors discussed that lean management leads to better energy consumption, waste, scrap, and pollution reduction, local sourcing, a better relationship with suppliers, cost-sharing, and information system integration. It also improves employee engagement and working environment and upskilling. Rosin et al. (2020) studied the link between Industry 4.0 and lean management. For Industry 4.0, they addressed the studies that considered autonomous robots, simulation, system integration, Internet of things (IoT), cloud, augmented reality (AR), big data, and cybersecurity, and from a lean perspective, Just in Time, Jidoka, Waste reduction, People and work, and foundations are considered. Few studies considered system integration, cloud, AR, and big data and from an optimization and control point of view. The link between IoT and lean is much more addressed in monitoring literature. Tortorella et al. (2020) studied designing the value stream in the context of Industry 4.0. They concluded that the cost and quality of technology adoptions and demand management require more research. For improving Takt-time, additive manufacturing, rapid prototyping, and 3D printing could be applied. Moreover, for implementing the continuous flows of production, collaborative robots, additive manufacturing, and 3D printing play an essential role. Varela et al. (2019) performed a survey and statistical analysis based on 252 questionnaires from industrial companies in Portugal and Spain. They confirmed the strong relationship between Industry 4.0 and the three pillars of sustainability. The correlation between Industry 4.0 and lean management is also confirmed in this study. Caldera et al. (2017) also conducted a literature review on the role of lean management on sustainability. They addressed the impacts on environmental performance with qualitative and quantitative evidence. For example, the 5S approach could decrease mixed wastes by 30%, and improving waste management and energy efficiency could be realized by a lean layout. Furthermore, effective scheduled maintenance and resource-saving via a value stream mapping approach could improve productivity (Chiarini 2014; Fliedner 2008). From a human resource perspective, employees' engagement and top management commitment to establish sustainablity culture are essential (Caldera et al. 2017). Romero et al. (2018) discussed the concept of cyber-physical technologies and lean production via discussing the different types of wastes. The authors also assessed Jidoka systems in the context of Industry 4.0 technologies. Table 4.1 shows some of these technologies, the impacts, and the implications on sustainability pillars.

The synthesis in the literature review shows that addressing the impacts of lean elements and Industry 4.0 and their interaction on life cycle cost is essential and has not received much attention in the literature. Table 4.2 summarizes some

Table 4.1 Industry 4.0 technologies, the outcomes on production process and planning, and the relation to the sustainability

Industry 4.0 technologies	Outcomes	Environmental benefits	Economic benefits	Social benefits
Smart control and self-adapting in real time	Reducing waiting times	•	•	
Automated guided vehicle	Reduce time and motion in material handling Reducing accidents	•	•	•
3D printing	Fast production Reducing wastes during the manufacturing process Avoiding overproduction	•	•	
Digital Kanban	Managing inventory level	•	•	
RFID	Automating inventory Reducing errors in materials and parts (inbound logistics optimization)	•	•	
Wearable computing	Reducing operator's workload Motion optimization		•	
Visual controls/ tracking operations	Tracking the operation/ real-time controlling Reducing downtime Improving maintenance	•	•	•
VR/AR	Reducing accidents Improving ergonomic posture Motion optimization Improving training			•

Table 4.2 The contributions of lean and Industry 4.0 in life cycle costs elements

Costs source	Lean impacts	Industry 4.0 impacts
Cost of manufacturing	Improving resource productivity Improving quality, reduction of wastes and defects, reducing leakage, reducing inventory costs Reducing lead time and setup time	Self-optimization manufacturing systems, improve equipment efficiency via rapid data sharing Reducing inventory costs with IoT, Optimizing the manufacturing process with additive manufacturing (Pagliosa et al. 2019, Tortorella et al. 2020)
Cost of eco-design	Improving the innovative ideas for sustainability, effective teamwork	Minimize the impacts of design changes, rapid prototyping and 3D printing, additive manufacturing (Pagliosa et al. 2019)
Transportation cost	Improving the layout design, reducing unnecessary transport	Routing optimization, smart platforms, using machine learning and AI and real-time navigation Transport optimization with data-driver enabled platform and IoT sensors-based connectivity
Training cost	Up-skilling, positive working environment (Piercy and Rich 2015), clear practices and procedures	Availability of different training, AR/VR-enabled training, learning factory concept, and life-long learning opportunities; perform complex tasks with high accuracy with robots; perform operational training (Pagliosa et al. 2019)
Energy costs	Reducing idle time with JIT and TPM (Piercy and Rich 2015)	Context-aware architecture for smart energy usage
Maintenance cost and downtime hours	VSM for maintenance operation, standardization of maintenance tasks, application of 5S and Kaizen in the maintenance process	Perform remote maintenance management of complex equipment, optimizing preventive maintenance with shared platforms (Pagliosa et al. 2019, Keivanpour 2021)
The economic value of the products	Reducing the leadtime, customer satisfaction	Optimize customer relationship management, personalized products, and modules
Cost of products recovery at EoL	Using VSM for optimizing disassembly process, standardization of disassembly process, minimizing changing working zones (Dayi et al. 2016), improving layout design of recovery facilities (Sundin et al. 2011)	AR/VR for reducing disassembly time, advanced simulation for improving disassembly sequence, improving recovery logistics (Keivanpour 2021)

implications of lean and Industry 4.0 on life cycle cost elements based on the literature review.

4.3   The Proposed Approach

The total cost of ownership models usually focuses on the cost of manufacturing and
operation and use deterministic parameters. The probabilistic models provide a com-
plete model, and they are more close to reality. Some probabilistic TCO models are
proposed in the context of energy policy for testing the impacts of novel technologies
(Danielis et al. 2018; Wu et al. 2015). In this study, in addition to the manufacturing
and operation, the cost of eco-design, product recovery, and end of life are also considered. It has been assumed that the manufacturer is responsible for products take-
back at the end of life (Fig. 4.3). The TCO model in (Dornfeld 2012) is considered as
the basic model and the other elements are integrated. The total cost of manufacturing,
operation, eco-design, lean, and Industry 4.0 technology investment as well as the cost
of end of life is divided by the value of the products and recovered value from parts
and materials (Eqs. 4.1–4.6). Future costs are discounted to the present value. Then a
Monte Carlo simulation is used to consider the stochastic elements. The distribution
of stochastic variables is shown in Table 4.3. The beta distribution and normal distribution are considered. In this chapter, 100,000 simulations are considered during the
whole life cycle of products. This simulation provides the statistical distribution of

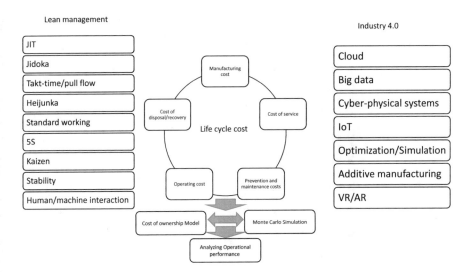

Fig. 4.3 The proposed framework

Table 4.3 The variables and the examples of the distribution functions

Stochastic variables of technology and practice adoption		Stochastic variables for eco-design		Stochastic variables of operation		Stochastic variables of recovery operation		Stochastic variables of recovery values	
$C_{Inv-4.0}$	Beta distribution	$C_{mtr.Imp}$	Normal distribution	$C_{Main-cr}$	Beta distribution	$C_{landfill}$	Normal distribution	V_{Recy}	Normal distribution
$C_{Inv-Lean}$	Beta distribution	$C_{diss.Imp}$	Normal distribution	H_{down}	Beta distribution	C_{T-EOL}	Normal distribution	V_{diss}	Normal distribution
$C_{opr-4.0}$	Beta distribution	$C_{recy.Imp}$	Normal distribution	C_{Enr}	Beta distribution	C_{diss}	Normal distribution		
$C_{opr-Lean}$	Beta distribution					C_{recy}	Normal distribution		

output as well as the possibility to analyze the probability of Industry 4.0 and lean management adoption and the impacts on TCO. Moreover, the sensitivity analysis of 100,000 outcomes based on the distribution of the input variables is possible.

$$TCO = \frac{\left(C_{mn} + C_{eco} + C_{set-up} + C_{Inv-4.0} + C_{Inv-Lean} + \dfrac{C_{opr}}{(1+i)^T} + \dfrac{C_{Rec}}{(1+i)^{T_{Eol}}} + \dfrac{C_{opr-4.0} + C_{opr-Lean}}{(1+i)^T} \right)}{n \times V_{p+} \left(nn_2 \times V_{Recov} \right)} \quad (4.1)$$

$$C_{set-up} = C_I + C_T + C_{train} \quad (4.2)$$

$$C_{opr} = C_{env} + C_{Enr} + \Sigma C_{cons} + H_{down} \times C_{Main_{cr}} + C_{Main\_pr} \quad (4.3)$$

$$C_{eco} = C_{mtr.Imp} + C_{diss.Imp} + C_{recy.Imp} \quad (4.4)$$

$$C_{Rec} = C_{landfill} + C_{T-EOL} + C_{diss} + C_{recy} \quad (4.5)$$

$$V_{Recov} = V_{Recy} + V_{diss} \quad (4.6)$$

4.4   Numerical Example

For testing the applicability of the proposed model, an experimental analysis is designed. In this experience, two factors including the operational excellence practices and their implications are considered. For the first factor, basic scenario, lean, Industry 4.0, and the combination of lean and Industry 4.0 are considered. For the second factor, the implications, three states of pessimistic, most probable, and optimistic are considered. Hence, ten experiments are designed. The implications are extracted based on the literature review (quantitative and qualitative studies), and the impacts on the cost elements presented in Table 4.2. The results are presented in Figs. 4.4, 4.5, and 4.6. As shown in Fig. 4.6, the combined application of lean and Industry 4.0 could improve TCO by 44, 49, and 53%, respectively.

4.5   Conclusion

The study on lean, Industry 4.0, and circular economy is emerging in the recent literature. However, few studies considered interaction among them from an operational performance perspective. This study proposed a probabilistic simulation model for considering the interaction of lean, Industry 4.0, and circular economy. This simulation could be used as a decision tool for analyzing the different impacts of operational excellence and technology revolution. It also aids in policymaking and highlights the challenges from an economic perspective. The operational

Definition of parameters

TCO	Total cost of ownership	C_{train}	Total training cost
C_{mn}	Cost of manufacturing	C_{env}	Ecological compensation
$C_{Inv-4.0}$	Investment cost for Industry 4.0 adoption	C_{Enr}	Energy costs
$C_{Inv-Lean}$	Investment cost for lean management	C_{Main_cr}	Corrective maintenance cost
$C_{opr-4.0}$	Operation cost for Industry 4.0 adoption	C_{Main_pr}	Preventive maintenance cost
$C_{opr-Lean}$	Operation cost for lean management	C_{Rec}	Cost of products recovery at EoL
C_{set-up}	Setup cost	$C_{landfill}$	Costs of landfill
C_{opr}	Cost of operation	C_{T-EOL}	Transportation cost-EoL
C_{eco}	Cost of eco-design	C_{diss}	Cost of disassembly
$C_{mtr.Imp}$	Cost of improving material	C_{recy}	Cost of recycling
$C_{diss.Imp}$	Cost of improving disassembly	H_{down}	Downtime hours
$C_{recy.Imp}$	Cost of improving recyclability	C_{cons}	Cost of consumables
C_I	Installation cost	i	Interest rate
C_T	Transportation cost	T	Time
T_{Eol}	The time that product reaches to EoL	V_{diss}	Economic value of the recovered parts
V_p	Economic value of the product	n	Number of products
V_{Recov}	Economic value of the recovery	n_2	Percentage of tack-back
V_{Recy}	Economic value of the recycled material		

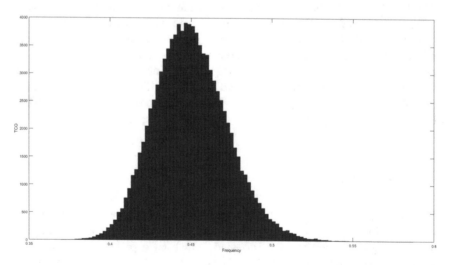

Fig. 4.4 TCO in the basic scenario

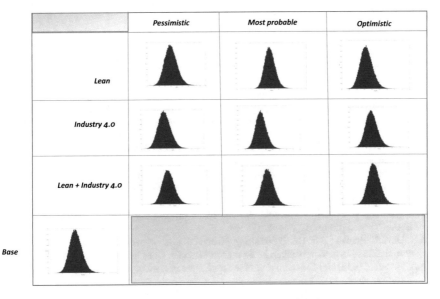

Fig. 4.5 The results of Monte Carlo simulation in a partial factorial design

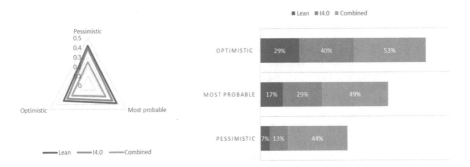

Fig. 4.6 The mean of TCO and the resulted improvement in TCO

performance of new technology adoptions and excellence operations in the circular economy is promising. In this study, the implications on the cost elements are extracted from the literature review and assigned based on experts' opinions in three levels of pessimistic, most probable, and optimistic. However, using primary data from real case studies is proposed as future research. Considering the other types of distribution functions for stochastics parameters and integrating fuzzy numbers are also suggested as future research.

References

C. Cagnetti et al., Lean production and Industry 4.0: Strategy/management or technique/implementation? A systematic literature review. Procedia Computer Sci. **180**, 404–413 (2021)

H.T.S. Caldera, C. Desha, L. Dawes, Exploring the role of lean thinking in sustainable business practice: A systematic literature review. J. Clean. Prod. **167**, 1546–1565 (2017)

A. Chiarini, Sustainable manufacturing-greening processes using specific lean production tools: An empirical observation from European motorcycle component manufacturers. J. Clean. Prod. **85**, 226–233 (2014)

A. Courtois, C. Martin-Bonnefous, M. Pillet, M. Pillet, *Gestion de production*. Les Ed. d'Organisation (2011)

R. Danielis, M. Giansoldati, L. Rotaris, A probabilistic total cost of ownership model to evaluate the current and future prospects of electric cars uptake in Italy. Energy Policy **119**, 268–281 (2018)

O. Dayi, A. Afsharzadeh, C. Mascle, A Lean based process planning for aircraft disassembly. IFAC-PapersOnLine **49**(2), 54–59 (2016)

N.K. Dev, R. Shankar, F.H. Qaiser, Industry 4.0 and circular economy: Operational excellence for sustainable reverse supply chain performance. Resourc. Conserv. Recycl **153**, 104583 (2020)

D.A. Dornfeld (ed.), *Green Manufacturing: Fundamentals and Applications* (Springer Science & Business Media, 2012)

B. Durakovic et al., Lean manufacturing: Trends and implementation issues. Period. Eng. Nat. Sci. (PEN) **6**(1), 130–143 (2018)

G. Fliedner, Sustainability: a new Lean principle, in Proceedings of the 39th Annual Meeting of the Decision Sciences Institute. 2008

S. Keivanpour, Toward joint application of fuzzy systems and augmented reality in aircraft disassembly. Cengiz Kahraman Serhat Aydın. Intelligent and Fuzzy Techniques in Aviation 4.0.: 20 (Springer, 2021)

M. Kurdve, M. Bellgran, Green lean operationalisation of the circular economy concept on production shop floor level. J. Clean. Prod. **278**, 123223 (2021)

M. Pagliosa, G. Tortorella, J.C.E. Ferreira, Industry 4.0 and lean manufacturing, J. Manuf. Technol. Manag. (2019)

N. Piercy, N. Rich, The relationship between lean operations and sustainable operations, Int. J. Oper. Prod. Manag. (2015)

D. Romero, P. Gaiardelli, D. Powell, T. Wuest, M. Thürer, Digital lean cyber-physical production systems: The emergence of digital lean manufacturing and the significance of digital waste, in *IFIP International Conference on Advances in Production Management Systems*, (Springer, Cham, 2018), pp. 11–20

F. Rosin et al., Impacts of Industry 4.0 technologies on Lean principles. Int. J. Prod. Res. **58**(6), 1644–1661 (2020)

E. Sundin et al., Improving the layout of recycling centres by use of lean production principles. Waste Manag. **31**(6), 1121–1132 (2011)

G.L. Tortorella et al., Designing Lean value streams in the fourth industrial revolution era: Proposition of technology-integrated guidelines. Int. J. Prod. Res. **58**(16), 5020–5033 (2020)

L. Varela et al., Evaluation of the relation between Lean manufacturing, Industry 4.0, and sustainability. Sustainability **11**(5), 1439 (2019)

G. Wu, A. Inderbitzin, C. Bening, Total cost of ownership of electric vehicles compared to conventional vehicles: A probabilistic analysis and projection across market segments. Energy Policy **80**, 196–214 (2015)

Chapter 5
Perspectives of Applications of Industry 4.0-Enabled Eco-Design in the Aircraft Industry

5.1 Introduction

The aircraft manufacturers have started integrating the environmental assessment into the product life cycle. Eco-design in the 3R approach is considered the most preferable practice in the context of the corporate social responsibility of manufacturers. Toxicology of the materials, energy consumption (particularly for aircraft engines), reducing nonrenewable energy resources, and noise reductions are among the common practices that are considered by OEMs. Several challenges of the application of DfE for complex products are addressed in the literature. The integration of different technical, functional, and environmental data during the life cycle of products, the complexity of analyzing the eco-efficiency models considering high numbers of parts and components in BoM of these products, and the lack of a systematic approach for integrating several design tools are some of these challenges (Keivanpour and Ait Kadi 2018).

Despite several methods and tools for eco-design, these tools are not fully adopted by aircraft manufacturers. The eco-design tools should be customized based on the requirements of the designers and considering multiple stakeholders in the value chain of the aircraft life cycle. A systematic approach is required for developing an eco-design tool for an aircraft manufacturer. For every systematic approach, the following steps are required:

- Identifying the problem
- Analyzing the problem
- Gathering information
- Identifying possible solutions to the problem
- Evaluating possible solutions
- Developing an action plan to correct the problem
- Revising the solutions

For implementing this systematic approach, we need to look at the problem from the perspectives of all key players. The values for each player considering different solutions should be assessed. Moreover, the long life cycle of the aircraft should be considered. One important challenge is data accessibility and tracing all the operation activities during the use phase. Hence, data analytics, horizontal integration, integrating the different product data management systems, and developing digital models are essential. Industry 4.0 facilitates the adoption of eco-design tools and aids in removing some existing challenges of applying eco-design tools. In this chapter, we focus on existing eco-design practices in the aircraft industry. We discuss the needed features and functionalities for developing DfE tools. DfE for maintenance operation and design for EoL are discussed in detail. Then, the implications of Industry 4.0 for improving the sustainability in these two stages in the life cycle of an aircraft are provided. The rest of the chapter is organized as follows: Sect. 5.2 reviews eco-design practices in the industry and discusses the challenges of applying these methods and tools in the industry. Section 5.3 discusses the implications of Industry 4.0 on design for maintenance and disassembly and Sect. 5.4 provides a summary.

5.2 Literature Review

In this section, a review of DfE in the aircraft industry including the existing approaches, methods, tools, and challenges for their implementation is provided. Moreover, the new trends in the studied of DfE in this industry and new strategies particularly for complex and high-tech products are discussed.

5.2.1 DfE in the Aircraft Industry, Tools, and Methods

Bollhöfer et al. (2012) discussed the green design in the aviation industry. They explained that despite the small share of air transportation in global emission (2%), the aircraft manufacturers, OEMs, and industry actors look for green practices to develop sustainable solutions. Long service life (30 years), as well as design and production cycle, encourage the key players in the industry to find green solutions before facing some compulsory legislations. The authors explained that the research field of aircraft eco-design is not well explored by academics and practitioners. The European "Clean Sky" program aims to investigate different eco-design requirements and develop the guidelines, methods, and tools in this context. Noise and emissions are two essential concerns of the growing air traffic. Hence, the main eco practices in the industry focused on reducing these impacts. Different goals in the Clean Sky program include green regional aircraft, smart fixed-wing aircraft, green rotorcraft, sustainable and green engines, systems for green operation, and eco-design. In designing the aircraft, optimization models for minimizing costs and weight play an essential role. The requirements of the regulations of REACH (The European Union (EU) Registration, Evaluation, Authorisation, and Restriction of Chemicals) and RoHS (Regulation and the Restriction of Hazardous Substances) are the other general practices in the aviation industry for phasing out the hazardous materials. Figure 5.1 shows the timeline of REACH based on EU Regulation (EC) No 1907/2006 (REACH-EU n.d.). If we consider the adoption of the REACH regulations in the market and the presence of all the other key actors, different strategies could be addressed in the context of the game theory. Figure 5.2 shows these options. The impacts of the sustainability decisions should be addressed in different horizon plans (short, medium, and long term). The aircraft manufacturers sooner or later should comply with the regulations. Hence, with sooner adoption, they can benefit from first-mover advantages and create more value for the stakeholders. Certain guidelines and environmental standards are used for evaluating the suppliers' performance. The material selection by designers should be supported by user-friendly decision-making tools. The LCA data should be integrated into other trade-off analysis methods for weight and costs. Bollhöfer et al. (2012) also compared three tools for eco-design (MCDM, S-LCA, and EFQM models) based on certain criteria such as flexibility, integrating into corporate systems, assessing the alternative options, considering product life cycle, combining operational and strategic decisions, considering the stakeholders, and sensitivity for the social criteria. Among these decision support tools, EFQM and MCDM have more potential for considering the integrated value chain. Lohner et al. (2011) discussed the eco-efficient material selection for Airbus and its suppliers as a part of the integrated technology demonstrator of the Clean Sky program. The authors proposed a quantitative life cycle assessment for the eco-efficiency of the material.

Lemagnen et al. (2009) discussed the management of hazardous material in the design of aeronautic parts based on the existing directives. They explained the important role of data in evaluating the hazardous risks in material, mixtures, and production processes. The names, labels, design specifications, suppliers, and risks

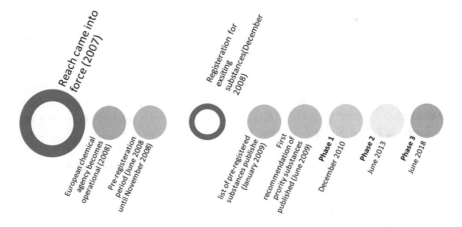

Fig. 5.1 The key dates in the REACH regulation

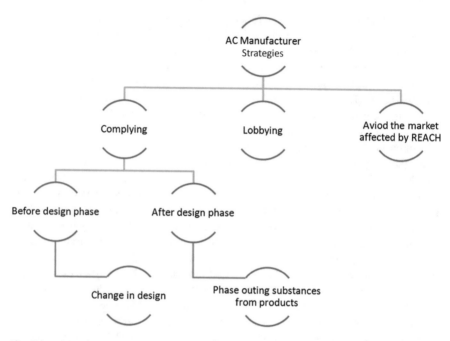

Fig. 5.2 Different strategies for adopting REACH directive in the context of Game theory

should be identified. The data could be extracted from BoM and product MIS. They proposed a systematic method for creating a compliance matrix based on the risk evaluation and the design recommendations. The authors explained that the calculation of hazard risk should be extended to the life cycle of the product, and the reliability of data and the input parameters can reduce the uncertainties in the results. In another study, Brissaud et al. (2008) discussed implementing REACH directives in engineering design. They explained that clustering of the substances is essential to the process of implementing the directives. The substance could be classified into three clusters: substances subject to authorization, substances of high concern, and List C. They addressed the required data during the life cycle of the products including materials database, databases of the processes of elaboration and handling in production, maintenance in the use phase, and the database of dismantling in the EoL phase. The risk evaluation methods could be integrated into existing design tools to develop a design support system. Moreover, this design support system should have data exchanges with ERP, PLM, CAD systems in engineering design.

In the majority of developed tools in this context, the integration of LCA and LCC with the other design tools is discussed. Ilg (2013) proposed an eco-design tool in the aviation industry. The concept of this design tool is based on developing an interactive interface for designers to facilitate using LCA data. In this tool, a type of aircraft in a decision dashboard could be selected. Then, the target subsystem, module, and component could be selected to present the environmental performance in graphical forms. Moreira et al. (2014) discussed the integration of the eco-design tool with product life cycle management. They focused on the textile material in aircraft construction and explained the opportunities and challenges of recycling textile (around 1% of total aircraft weight) considering the design and PLM systems. Moawad (2019) developed a stochastic eco-efficiency approach to the eco-design of additive manufacturing in the aircraft industry. The author integrated LCA data and LCC and use them in an eco-efficiency diagram for determining the most preferable eco-friendly alternative. Then, a sensitivity analysis based on certain scenarios (LCC, LCA, and eco-efficiency) is performed. Calado et al. (2019) also used integration of LCA and LCC in the composite material selection of aircraft structure.

Balança et al. (2014) discussed the design for EoL in high-tech products. They emphasized the characteristics of high-tech products as an important factor in defining design for EoL strategies. The complexity of the architecture, the long life cycle, and the design process are some of these features. They explained identification versus integration for the strategies. For identifications, the EoL indicators should be determined via assessment methods; the rules and principles should be addressed to develop guidelines. For the integration of the strategies, the implementing guidelines could be led to measure indicators and modifying and developing relevant assessment methods. They also discussed that the visibility of EoL industries and good communication between designers, manufactures, and EoL initiatives are essential. The communication between actors improves the product recovery processes and enhances the integration of design strategies. The technology readiness level also plays an important role. Keivanpour et al. (2014) proposed a model for integrating the value network of stakeholders in design for EoL in the aircraft industry. They classified eco practices related to EoL and discussed the principles of considering the stakeholders in the decision-making process (see Figs. 5.3 and 5.4). They proposed a guideline with six steps for integrating the stakeholders in selecting eco-design practices (see Fig. 5.5).

Sabaghi et al. (2016) proposed a decision support tool with a fuzzy inference technique for eco-design in the aircraft industry. They defined five levels of criteria in the developed MCDM for addressing environmental, economic, and social indicators.

Hence, the goal in DfE tools in the aircraft industry is developing a user-friendly and interactive interface for integrating design tools, LCA/LCC, and other product databases considering the values for all stakeholders (see Fig. 5.6). In ranking and selection of eco-design practices, several attributes including technical, environmental, and trade-off analysis should be considered (Fig. 5.7).

5.2.2   The New Trends in DfE for the Complex Products

The new trends in DfE focused on eco-design of PSS, value, and data-driven approaches, integrating visualization and mapping capacity in design tools, and using data mining and AI techniques for enhancing decision-making. Here, a review of recent publications is provided.

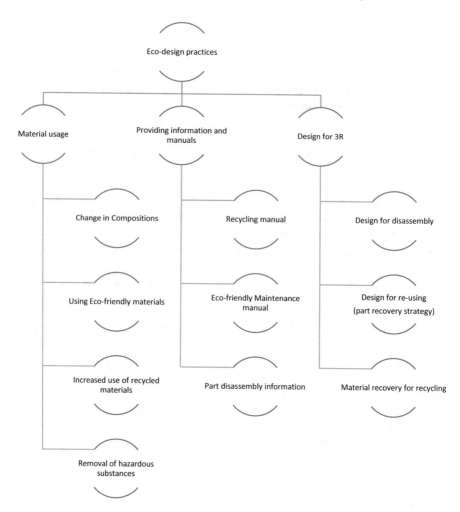

Fig. 5.3 The example of eco-design practices for improving EoL of aircraft (adapted from Keivanpour et al. 2014)

Castagne et al. (2009) used value-driven optimization in the design of aircraft fuselage panels. They considered the manufacturer's profit, the costs during the use phase (considering the revenue of airlines), and the surplus-value related to the competitor's products. They considered material, fabrication, and assembly costs. They stressed that the value-driven approach is an effective method for considering the values of stakeholders. Bertoni et al. (2013) developed a value visualization approach in designing PSS. They used color codes for mapping the value in the CAD 3D model to visualize the value information for the designers. Bertoni et al. (2016) also discussed the use of a value-driven approach in the design of PSS. They discussed that the decision design of PSS should include knowledge and product value stream. The value-driven approach could integrate these two elements and

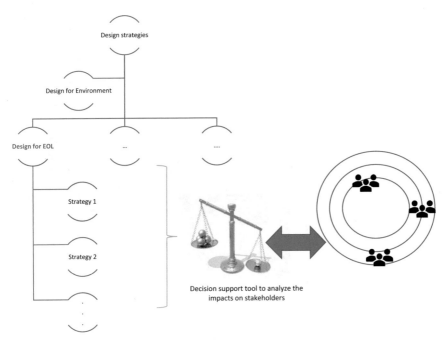

Fig. 5.4 Decision support tool for DfE considering the stakeholders

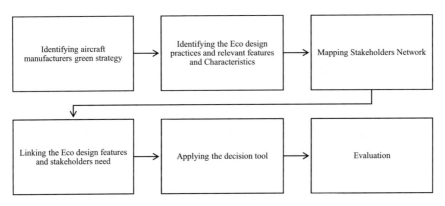

Fig. 5.5 The six-step approach for integrating stakeholders value in DfE (adapted from Keivanpour et al. 2014)

facilitate the integration of qualitative and quantitative criteria for designers. Bertoni et al. (2020) applied a value-driven approach and machine learning in sustainability design and provide a case study in the aerospace industry. In the sustainability model, they considered sustainability criteria based on LCA data. For the value model, they considered quantitative criteria such as customer revenue model, maintenance, and manufacturing costs. For qualitative criteria, serviceability, commonality, and scalability are integrated into the model. All the criteria from sustainability

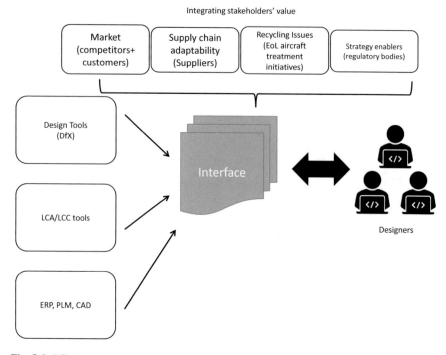

Fig. 5.6 DfE interface

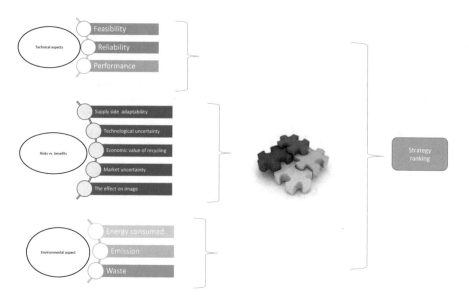

Fig. 5.7 Multiattribute analysis in DfE

and value models will be integrated into a digital model, and then a machine learning approach is used for selecting the best design scenario and visualization into CAD 3D model for the designer.

Keivanpour and Ait Kadi (2018) discussed the role of visualization in the eco-design of complex products. The authors proposed a data mining approach and tree map for mapping the ecoefficiency of the parts in BoM of a complex product. They discussed the existing challenges in DfE techniques for complex products and the role of data mining and visualization in designing a systematic approach for DfE (see Fig. 5.8). The author used tree map and color codes for addressing the ecoefficiency profile of the complex products (See Fig. 5.9).

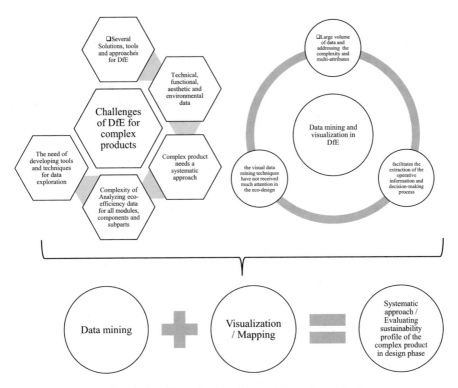

Fig. 5.8 Existing challenges in DfE and the role of data mining and visualization

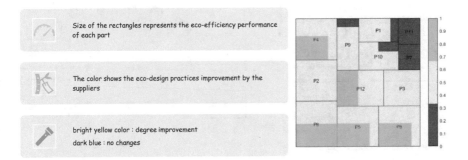

Fig. 5.9 Using color codes and mapping in addressing the sustainability profile of a complex product (adapted from Keivanpour and Ait Kadi 2018)

Saves (2021) discussed the multidisciplinary nature of aircraft design and the role of the advanced optimization and simulation model in considering different design parameters.

The synthesis in the literature highlight the following points:

- Integrating different product management databases, LCA data, and LCC are essential in developing DfE tools.
- Developing a user-friendly interface for designers with visualization capacity could facilitate the integration of qualitative and quantitative criteria.
- DfE for PSS products is a new trend in the aircraft industry that value-driven approach is appropriate in this context.
- Considering the need for analyzing the large volume of data during the life cycle of the complex products, data-driven approaches, and data analytics tools are essential.
- Integrating DfE interface with designers' tools (3D CAD) model via data- and value-driven approach could facilitate the adoption of DfE tools for designers.
- Applications of the Industry 4.0 technologies in the context of DfE in the aircraft industry is a fresh topic, and more research is required to address the opportunities and challenges.

5.3   Perspectives of Applications of Industry 4.0 in DfE of Aircraft

The simple life cycle of the aircraft and the main processes are shown in Fig. 5.10. Raw materials, fuel, water, energy consumption, and transport are the main sources of environmental impacts. In this chapter, we focus on two main processes for addressing the implications of Industry 4.0 technologies on enhancing sustainability design: maintenance and disassembly.

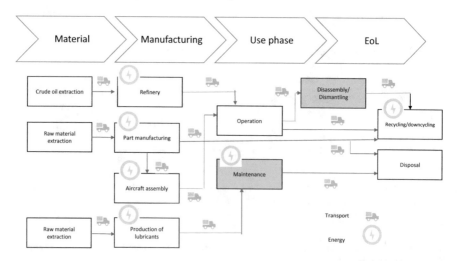

Fig. 5.10 Aircraft life cycle

5.3.1 DfE and Sustainable Maintenance Operation

Design for the sustainability of maintenance operation is essential for aircraft manufacturers, as the operation and maintenance contribute to the main environmental impacts in the whole life cycle of the aircraft. According to EPA/310-R-97-001 (n.d.), three factors are essential in assessing the environmental impacts of aircraft in the maintenance phase: (1) hours of flight time, (2) number of landing and take-off cycles, and (3) calendar length of time from prior maintenance. Chester (2008) also used EPA/310-R-97-001 category, to evaluate the impacts of the maintenance phase in LCA of aircraft (Keivanpour and Kadi 2015). The main maintenance activities based on the literature review are shown in Fig. 5.11. Evaluating the impacts of some of these activities is challenging and depends on the different airlines' operations. Sometimes the data are not available. For example, for lubrication and cleaning, the data related to the amount of used oils and greases, type and amount of the

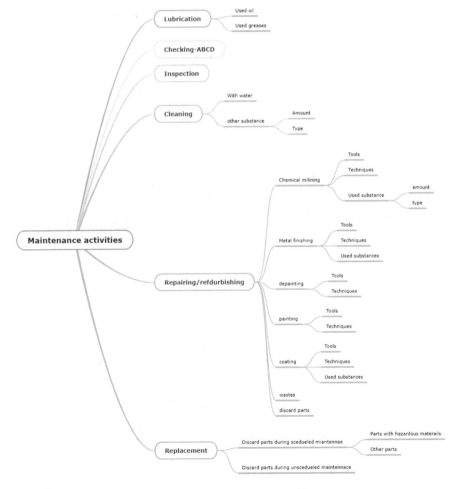

Fig. 5.11 Aircraft maintenance activities and the environmental impacts

cleaners, and the used substances should be processed to provide effective design feedbacks. Vertical and horizontal integration in the supply chain provide transparency in data and aid designers to access the data during the operation and maintenance phase.

The other maintenance activities that contribute considerably to the environmental impact of the aircraft are repairing, refurbishing, and parts replacements. Different repairing processes such as chemical mining, painting, coating, metal finishing should be evaluated in terms of the required tools, the used substances, and techniques. The discard parts including the parts with hazardous substances and non-repairable items during scheduled and unscheduled maintenance should be traced effectively. Accessing large databases and data processing should be facilitated for decision-making and providing valuable information in designers' dashboards. Figure 5.12 shows a preliminary architecture for processing the maintenance data for DfE.

The proposed approach has seven elements. (1): The first part is the management of hazardous materials. In addition to applying the best practices in storage, disposal, and transport, identifying and tracing these risks are essential. According to Junior et al. (2018), automation and control by computers could reduce the safety risks. The objective is to reduce, eliminate or substitute hazardous materials in the design process. If eliminating is not possible, the visualized manual for

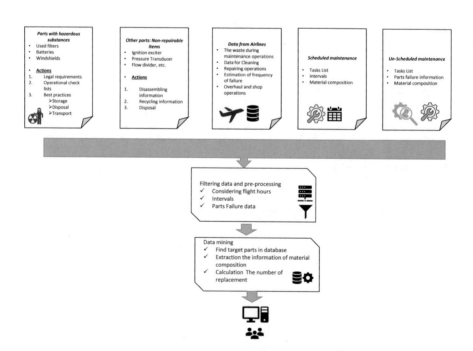

Fig. 5.12 Data-driven approach for design for the sustainability of maintenance activities

maintenance operation or using AR/VR for training the operators can reduce the safety risks. (2): The second element is the management of non-repairable items. All the data related to these items should be collected and processed during the maintenance operation. The decision support system is required to compare the recycling/disposal trade-off analysis. (3): The other category of the required data is the airline operations including the cleaning processes, the estimation of the frequency of failures, overhaul, and shop visit operations. (4,5): The data of maintenance operations in two categories of scheduled and unscheduled maintenance should be collected and analyzed. For the scheduled maintenance, it includes tasks list, interval, and the material composition of replaceable items. For unscheduled maintenance, the parts failure information is also required. (6): All of these data should be cleaned and preprocessed at the design stage considering the flight hours, intervals of maintenance operations, and parts failure data. (7): Then, data mining approaches could be applied to determine the target parts and clustering the parts based on the number of replacements, material composition, recycling/disassembly, and disposal trade-off analysis, and safety concerns. These data analytics approach aids in developing a holistic approach for evaluation of environmental impacts during maintenance operation and for an effective design for sustainable maintenance. Table 5.1 highlights the challenges in sustainable maintenance operations and Industry 4.0 implications.

Table 5.1   The challenges in sustainable maintenance design and the Industry 4.0 contributions

Challenges	Industry 4.0 possible solutions
Access to the data of wastes during the maintenance operation	Using horizontal integration for accessing the data during operations
Data regarding the types of cleaning solvent	Tracking the maintenance impacts of each subsystem concerning the relevant challenges
	Using the cloud-based system for sharing the operation data between key stakeholders
Evaluation of the impacts of repairing operations due to: Increased probability of failures after each repair The success rate of repairing operation Estimation of the frequency of failure of the components and subparts (different approaches and considering the probabilistic nature of the failure rate of components)	Using AI tools including ML, fuzzy logic, and decision tree for pattern recognition of failures after repairing operation
Considering repairing operation includes cleaning, chemical milling, metal finishing, de-painting, coating, and subparts replacements in LCA	Using horizontal integration for accessing the data during operations Tracking all maintenance operations
Overhaul and shop visit operations—The tools and techniques and used materials and substances during these activities	Using digital twins, AR/VR technology in simulation and analyzing the different overhaul and shop visit operations

5.3.2   Design for Disassembly

In this section, a conceptual framework is proposed for DfD.4.0 (see Fig. 5.13). The proposed architecture has four building blocks based on the synthesis in the literature and the existing gaps in DfE of complex products and the new trends. As discussed in Sect. 5.2.2, the value-driven approach provides the opportunity to consider several qualitative and quantitative factors at the design stage. For DfD, there are different criteria in ease of disassembly, the time for disassembly operation, and the performance. For the performance, in addition to the costs, the disassembly rate and the other EoL alternatives should be evaluated. In addition, the changes in structure should be evaluated from the perspectives of serviceability, scalability, and the commonality in used technology and architecture. The detailed criteria for disassembly are shown in Fig. 5.14. The second element is using AI tools such as fuzzy logic and machine learning considering the uncertainties of factors, the volume of data that should be analyzed, and the required rules that should be extracted from a value-driven approach. This analytical tool aids in integrated decision-making and facilitating the comparison of different design alternatives for the designers. The third element in this architecture is integrating the concept of co-creation. Integrating the different stakeholders' values is essential. The last element is using AR/VR for enhancing the design space. The process for creating an AR experiment and the steps for creating an AR model for a subassembly part is shown in Figs. 5.15 and 5.16 respectively. Using virtual environments for developing disassembly sequences has recently received attention from scholars (e.g., Liu et al. 2016; Zou et al. 2019). Di Gironimo et al. (2004) proposed a method for integrating visualization, 3D model, and digital mock-up for disassembly sequence planning in the virtual environment and improving operators' postures during the maintenance and disassembly operations. The proposed method has four steps. After creating the digital mock-up of the complex assembly, the system analysis is performed. In step 2, the disassembly sequences will be proposed. 3D visualization and CAD system will be integrated into this step for facilitating the analysis for designers. In step 3, the

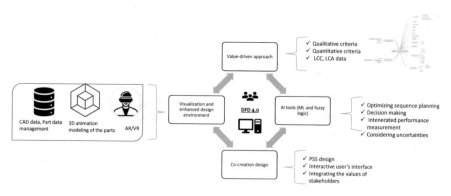

Fig. 5.13 The proposed architecture for DfD 4.0

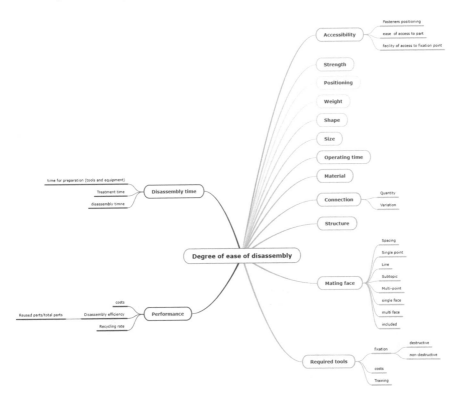

Fig. 5.14 DfD key parameters

accessibility and maintainability of the system will be performed. Finally, in step 4, via human modeling, the optimal postural sequence will be developed for the ergonomic optimization of the tasks. Bernard et al. (2020) compared different simulation tools including human modeling, VR, and physical mock-up in ergonomic optimization in the aviation industry. They concluded that VR plays an essential role in design for maintainability, and a new paradigm should be developed for focusing on different interrelated maintenance tasks during analysis and designing the virtual platform and 3D visualization for the design purposes. Bouyarmane et al. (2020) used CAD and human modeling for analyzing design for disassembly for improving ergonomic risks score and accessibility. The authors discussed that VR could facilitate the estimation of disassembly time. Frizziero et al. (2019) proposed an AR-based design for disassembly for a gearbox. They used different disassembly sequence planning and then evaluated the scenarios in an AR environment. The authors explained several advantages for this design approach including reducing the risks for the operators, reducing the design and development cycle and human errors, as well as improving the efficiency in time and costs of the production. They also mentioned some disadvantages that would be solved by technology advancements such as the required time for creating the animations for the complex products with many parts and the programming. FREDDI AND FRIZZIERO (2020) also proposed an

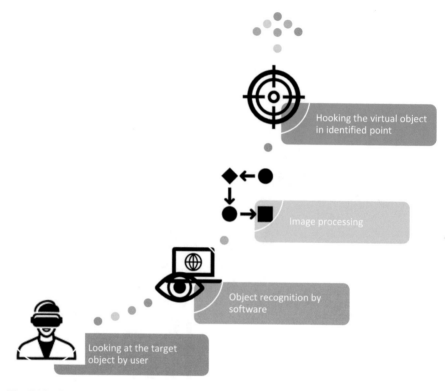

Fig. 5.15 The processes for creating an AR experiment

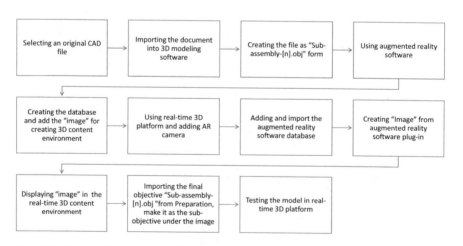

Fig. 5.16 The steps for creating an AR model for a subassembly part

AR case study for design for the disassembly of a tailstock. Different Industry 4.0 technologies are applied in the context of the circular economy. Rocca et al. (2020) discussed the application of VR and the digital twin in the circular economy via a

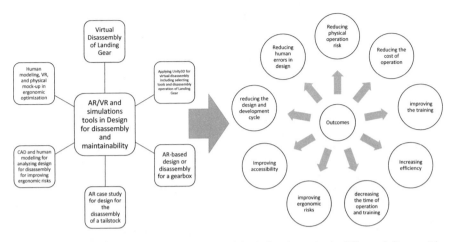

Fig. 5.17 Example of applications of Industry 4.0 in design for maintainability and disassembly and the outcomes

laboratory case. Figure 5.17 summarizes these applications and the outcomes based on the review of the papers in this section.

5.3.3   Designing an Interactive Decision Dashboard for DfE 4.0

In this section, a preliminary framework for developing a visualized decision dashboard for DfE in the context of Industry 4.0 is provided. The building blocks of the architecture are proposed based on the synthesis in the literature review in Sect. 5.2. As discussed earlier, there are different challenges in DfE of complex products. Several models and tools should be used at the design stage. Different databases including technical, environmental, and economic data should be integrated. Hence, visualization and data analytics aid the eco-design team in having a systematic and holistic approach to DfE. The developed decision tool integrates the massive data and advanced simulation, and optimization provide multivariant analysis. Hence, the first step is developing a checklist with the eco-design team to determine the level of maturity in design and the potential for adoption of digital technologies. Table 5.2 shows some questions that should be checked with the design team. This checklist aids in identifying the existing efforts, the gaps, and the priorities for the manufacturer. For avoiding complexity, developing a pilot study focusing on one module of complex products is proposed. The approach for developing the decision tool is shown in Fig. 5.18. By focusing on one component or module, the related stakeholders in the value chain could be identified. Different design scenarios should be developed based on design parameters, the sustainability objectives, and the long-term plan of the manufacturer. Then, the related criteria should be identified. These criteria include the economic, safety, environmental, and complexity in

Table 5.2 The checklist for the eco-design team

No	Descriptions	Level of adoption	Comments
1	Is there any systematic approach to eco-design?	● ○ ○ Low Medium High	
2	Is there any eco-design tool that integrates: • Product design and analysis • BoM, CAD model, product structure, the composition of materials • Calculate durability • Apply the changes: product structure, component merger, change of connection type, change of material, modification of shape and geometry, etc.	● ○ ○ Low Medium High	
3	Have you already integrated LCC and LCA into your eco-design tools?	● ○ ○ Low Medium High	
4	Do you use a visualization technique for eco-design?	● ○ ○ Low Medium High	
5	Are you using a value-driven approach?	● ○ ○ Low Medium High	
6	Do you integrate eco-design at the component level?	● ○ ○ Low Medium High	
7	Availability of data for LCC, LCA?	● ○ ○ Low Medium High	
8	How do you integrate the circular economy?	● ○ ○ Low Medium High	
9	Do you use the integrated database?	● ○ ○ Low Medium High	
10	How customers are involved at the design stage?	● ○ ○ Low Medium High	
11	How the other stakeholders are involved?	● ○ ○ Low Medium High	

(continued)

Table 5.2 (continued)

No	Descriptions	Level of adoption	Comments
12	Do you use a collaborative platform for design?	● ○ ○ Low Medium High	
13	Do you use cloud-based platforms for facilitating the integration of the stakeholders' values into the design stage?	● ○ ○ Low Medium High	
14	Do you use VR/AR at the design stage?	● ○ ○ Low Medium High	
15	Do you use additive manufacturing at the design stage?	● ○ ○ Low Medium High	
16	Do you have an integrated product life cycle management system?	● ○ ○ Low Medium High	
17	If the information in PLM will be integrated into the design stage?	● ◐ ○ Low Medium High	

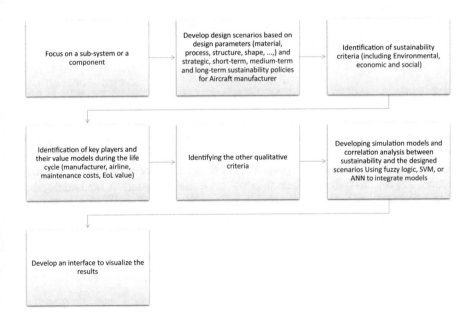

Fig. 5.18 The essential steps for developing a decision DfE dashboard

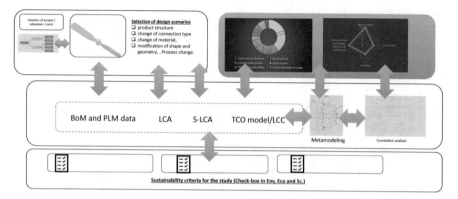

Fig. 5.19 Dashboard presenting the main elements of the eco-design tool

design and manufacturing. The simulation model should be developed for correlation analysis between the design scenarios and sustainability performance. The impacts on the value for different actors in the value chain should be analyzed. In this step, a fuzzy logic model, decision tree, or other AI tools could be applied for extracting the rules from the data. An interactive interface will be developed to use by designers in facilitating the integration of the design parameters. Figure 5.19 shows the decision dashboard architecture. A module or part could be selected from BoM and the design parameters based on structure, material, joint type, and the process will be determined. Different databases including PLM, LCA, LCC, and SLCA should be accessible and the integrated analysis could be used for developing metamodels and correlation analysis. The sustainability performance aids the design team to have a holistic approach to DfE.

5.4   Conclusion

In this chapter, the perspectives of the applications of Inustry4.0 in the aircraft industry are discussed. The discussions about the applications in design for sustainable maintenance and DfD are provided. For the maintenance, the role of data analytics, and for DfD, the applications of VR/AR are discussed. A methodological framework for developing an eco-design tool and the interactive decision dashboard are also presented. The application in a pilot case study is proposed as future research.

References

F. Balança, M. Lemagnen, S. Pompidou, N. Perry, Development of a design for end-of-life approach in a strongly guided design process. Application to high-tech products. In *Going Green–Care Innovation* (2014), pp. 1–9

F. Bernard, M. Zare, J.C. Sagot, R. Paquin, Using digital and physical simulation to focus on human factors and ergonomics in aviation maintainability. Hum. Factors **62**(1), 37–54 (2020)

A. Bertoni, M. Bertoni, O. Isaksson, Value visualization in product service systems preliminary design. J. Clean. Prod. **53**, 103–117 (2013)

A. Bertoni, M. Bertoni, M. Panarotto, C. Johansson, T.C. Larsson, Value-driven product service systems development: Methods and industrial applications. CIRP J. Manuf. Sci. Technol. **15**, 42–55 (2016)

A. Bertoni, S.I. Hallstedt, S.K. Dasari, P. Andersson, Integration of value and sustainability assessment in design space exploration by machine learning: An aerospace application. Design Science, 6 (2020)

E. Bollhöfer, T. Müller, J. Woidasky, EcoDesign 2020: Green design requirements and implementation in the aircraft industry (Fraunhofer Gesellschaft: Karlsruhe/Pfinztal, 2012), p. 202020. https://gin.confex.com/gin/2012/webprogram/Manuscript/Paper3624/EcoDesign

H.Bouyarmane, M. El Amine, M. Sallaou, Disassembly evaluation during the conceptual design phase to ensure a better valorisation of products at the end of life cycle disassembly evaluation during the conceptual design phase to ensure a better valorisation of products at the end of life cycle (2020)

D. Brissaud, M. Lemagnen, F. Mathieux, A new approach to implement the REACh directive in engineering design, in *DS 48: Proceedings DESIGN 2008, the 10th International Design Conference, Dubrovnik, Croatia* (2008), pp. 1335–1340

E.A. Calado, M. Leite, A. Silva, Integrating life cycle assessment (LCA) and life cycle costing (LCC) in the early phases of aircraft structural design: An elevator case study. Int. J. Life Cycle Assess. **24**(12), 2091–2110 (2019)

S. Castagne, R. Curran, P. Collopy, Implementation of value-driven optimisation for the design of aircraft fuselage panels. Int. J. Prod. Econ. **117**(2), 381–388 (2009)

Chester, *Life-cycle Environmental Inventory of Passenger Transportation in the United States* (Institute of Transportation Studies, UC Berkeley, 2008). http://escholarship.org/uc/item/7n29n303

G. Di Gironimo, G. Monacelli, S. Patalano, A design methodology for maintainability of automotive components in virtual environment, in *DS 32: Proceedings of DESIGN 2004, the 8th International Design Conference, Dubrovnik, Croatia* (2004)

EPA/310-R-97-001 (n.d.). https://cfpub.epa.gov/ols/catalog/advanced_full_record.cfm?&FIELD1=SUBJECT&INPUT1=Impacts&TYPE1=EXACT&LOGIC1=AND&COLL=&SORT_TYPE=MTIC&item_count=941&item_accn=207733

M. Freddi, L. Frizziero, Design for disassembly and augmented reality applied to a tailstock, in *Actuators* (vol. 9, no. 4, 2020). Multidisciplinary Digital Publishing Institute, p. 102

L. Frizziero, A. Liverani, G. Caligiana, G. Donnici, L. Chinaglia, Design for disassembly (DfD) and augmented reality (AR): Case study applied to a gearbox. Mach. Des. **7**(2), 29 (2019)

R. Ilg, Modeling complex aviation systems-the eco-design tool EcoSky, in *Green Design, Materials and Manufacturing Processes* (2013), pp. 293–296

J.A.G. Junior, C.M. Busso, S.C.O. Gobbo, H. Carreão, Making the links among environmental protection, process safety, and Industry 4.0. Process Saf. Environ. Prot. **117**, 372–382 (2018)

S. Keivanpour, D. Ait Kadi, Strategic eco-design map of the complex products: Toward visualisation of the design for environment. Int. J. Prod. Res. **56**(24), 7296–7312 (2018)

S. Keivanpour, D.A. Kadi, A sustainable approach to aircraft engine maintenance. IFAC-PapersOnLine **48**(3), 977–982 (2015)

S. Keivanpour, D.A. Kadi, C. Mascle, Toward a decision tool for eco-design strategy selection of aircraft manufacturers considering stakeholders value network. SAE Int. J. Mater. Manufac. **7**(1), 73–83 (2014)

M. Lemagnen, F. Mathieux, D. Brissaud, How managing more efficiently substances in the design process of industrial products? An example from the aeronautics sector. *arXiv preprint arXiv:0903.4071* (2009)

T. Liu, M. Chen, Y. Wang, Design and research of virtual disassembly system for aircraft landing gear, in *2nd International Conference on Computer Engineering, Information Science & Application Technology (ICCIA 2017)* (Atlantis Press, 2016)

H. Lohner, I. Delay-Saunders, K. Hesse, A. Martinet, M. Beneke, P. Kalyan, B. Langer Eco-efficient materials for aircraft application (No. 2011-01-2742). SAE Technical Paper (2011)

E. Moawad, *A Stochastic Eco-Efficiency Approach to Support the Eco-Design of Additive Manufactured Components in the Aircraft Industry* (Doctoral dissertation, Polytechnique Montréal, 2019)

N. Moreira, D. Aït-Kadi, D.R. Vieira, A. Romero, L.A. de Santa-Eulalia, Y. Wang, Integrating eco-design and PLM in the aviation completion industry: A case study, in *IFIP International Conference on Product Lifecycle Management*, (Springer, Berlin, 2014), pp. 169–180

REACH-EU (n.d.). https://ec.europa.eu/environment/chemicals/reach/reach_en.htm

R. Rocca, P. Rosa, C. Sassanelli, L. Fumagalli, S. Terzi, Integrating virtual reality and digital twin in circular economy practices: A laboratory application case. Sustainability **12**(6), 2286 (2020)

M. Sabaghi, C. Mascle, P. Baptiste, R. Rostamzadeh, Sustainability assessment using fuzzy-inference technique (SAFT): A methodology toward green products. Expert Syst. Appl. **56**, 69–79 (2016)

P. Saves High dimensional multidisciplinary design optimiza-tion with Gaussian process for eco-design aircraft (2021)

F. Zou, Y. Chen, K. Wei, Landing gear virtual disassembly and assembly system based on Unity3D, in *2019 IEEE 4th Advanced Information Technology, Electronic and Automation Control Conference (IAEAC)* (Vol. 1, IEEE, 2019), pp. 2745–2748

Index

Printed in the United States
by Baker & Taylor Publisher Services